Contemporary Discourse
in the Field of
ASTRONOMY ™

Scientific Information About the Universe and the Scientific Theories of the Evolution of the Universe

An Anthology of Current Thought

Edited by Rick Adair

The Rosen Publishing Group, Inc., New York

This book is dedicated, first, to my late father, Roy Adair, who fostered a love of science in me and showed me where to find M33 in the night sky. It is also dedicated to my wife, Susan Petty, and children, Zachary and Devin, whose forbearance, encouragement, and support made this possible. And, last but not least, to my editor, Leigh Ann Cobb, whose own forbearance and delightful disposition made this project a treat.

Published in 2006 by The Rosen Publishing Group, Inc.
29 East 21st Street, New York, NY 10010

Library of Congress Cataloging-in-Publication Data

Scientific information about the universe and the scientific theories of the evolution of the universe: an anthology of current thought/edited by Rick Adair.
    p. cm.—(Contemporary discourse in the field of astronomy)
Includes bibliographical references and index.
ISBN 1-4042-0397-4 (library binding: alk. paper)
1. Cosmology—Popular works. 2. Expanding universe—Popular works. 3. Dark energy (Astronomy)—Popular works. 4. Dark matter (Astronomy)—Popular works.
I. Adair, Rick. II. Title. III. Series.

QB982.S39 2005
523.1—c22

                                        2004027659

Manufactured in the United States of America

On the cover: Bottom right: The Hubble Ultra Deep Field (HUDF) is the deepest view of the universe to be achieved by humankind, revealing the first galaxies to emerge shortly after the big bang. Bottom left: Galileo Galilei. Center left: the Dumbell Nebula. Top right: solar flares.

CONTENTS

Introduction   **5**

1. The Dawning of a New Age   **11**

"Executive Summary" by the Board on Physics
   and Astronomy   12
"On the Age of the Universe" by Wendy L. Freedman   26
"The Greatest Story Ever Told" by Ron Cowen   37
"New Measurement of Stellar Fusion Makes Old
   Stars Even Older" by Kim Krieger   49
"The Myth of the Beginning of Time"
   by Gabriele Veneziano   52

2. Dark Energy and the Accelerating Universe   **72**

"Supernovae, Dark Energy, and the Accelerating
   Universe" by Saul Perlmutter   73
"Illuminating the Dark Universe" by Charles Seife   94
"The Cosmic Symphony" by Wayne Hu and
   Martin White   101
"Galaxy Clusters Bear Witness to Universal
   Speed-Up" by Adrian Cho   118
"Dark Doings" by Ron Cowen   121

3. Dark Matter   **132**

"The Search for Dark Matter" by David B. Cline   133
"The Hunt for Dark Matter in Galaxies"
   by Ken C. Freeman   145
"The Cosmic Web" by Robert A. Simcoe   151

4. Bringing the Universe Closer to Home   **169**

"Distant Elements of Surprise" by Max Pettini   170

"The Cradle of the Solar System" by J. Jeff Hester,
   Steven J. Desch, Kevin R. Healy, and
   Laurie A. Leshin   177
"Planet Signs? Sifting a Dusky Disk" by Ron Cowen   183

Web Sites   **186**
For Further Reading   **186**
Bibliography   **188**
Index   **190**

Introduction

Most of the articles gathered in this anthology were selected to capture the revolution that has shaken cosmology and our understanding of the universe in the last few years.

The pivotal year for this revolution was 1998. Up to that time, most astronomers were confident that the universe had been expanding for 10 billion to 20 billion years—since its birth after the big bang. The expansion started with inflation: a poorly understood growth spurt—over an amazingly short period of time—of an incredibly hot, much smaller universe. Some time later, the energy and matter in the universe coalesced into stars and galaxies. The afterglow of the big bang cooled along with the expanding universe and now fills the sky with a nearly uniform cosmic microwave background.

Four pillars of thought support this view. The first, and earliest, was Albert Einstein's general theory of relativity, published in 1915. This framework combines the three dimensions of space with time to show how the curvature of "space-time" is related to gravitational force. This means that everything from a neutron to a neutron star wraps space-time around it into a proportionately sized, four-dimensional dimple. The theory also implies that if the universe is expanding, then it is plausible that at some earlier time it was much smaller.

The next pillar is American astronomer Edwin Hubble's 1929 discovery that the universe is expanding. Expansion means that everything is accelerating away from everything else. In other words, the farther away something is, the faster it seems to speed away. To show this, Hubble relied on the light from variable brightness stars called Cepheids to estimate the distance and speed of each star. He estimated the distances much as we judge how far away a car is by the brightness of its headlights. Hubble compared the observed brightness of a Cepheid to its true brightness, which is indicated by the time it takes the Cepheid to flicker through a complete brightness cycle.

Hubble's speed estimates used redshift, which is the shift of a receding object's light toward lower frequencies (red is at the low end of the visible spectrum), in much the same way the tone of a train's whistle drops as the train moves away. The various elements in stars, such as hydrogen, emit light (and other electromagnetic radiation) at specific, well-known frequencies, making them excellent redshift markers.

Hubble found that redshifts increased with the distance of these stars, which meant that speed increased with distance. According to this discovery, the universe was expanding.

With this discovery, Einstein had to make a minor correction to his general theory of relativity, which originally had a constant term to conform to the then-widespread belief that stars maintained constant distances from each other. The constant term amounted to an "antigravity" force that kept the stars from eventually crashing together under gravitational attraction. But the discovery of expansion meant the equations didn't need the constant.

The third pillar supporting this view of the universe is big bang nucleosynthesis (BBN), a theory first published in 1948 by American physicist Ralph Alpher and Ukrainian physicist George Gamow about the creation of the light elements deuterium (a hydrogen isotope), helium, and lithium soon after the big bang. It explains why astronomers see approximately the same amounts of these light elements everywhere they look, suggesting that the elements formed together.

BBN theory assumes that the big bang picture is valid and that the universe was once very hot and dense. Under these conditions, a continuous bath of photons kept baryons (neutrons and protons) from combining. Finally, as the universe expanded and cooled, baryons combined through nuclear fusion to form elements. But only the lighter elements, made of a few protons and neutrons, were stable. This was because the universe was still extremely hot. This "nucleosynthesis" stopped after about three minutes, when the temperature and density fell below values nuclear fusion requires.

Two key features of BBN are that the amounts of these elements depend only on the number of photons per baryon present before fusion started and that the abundances of deuterium and helium after BBN stops aren't very sensitive to this photon-baryon ratio. BBN will yield a universe made up of about 25 percent helium relative to the mass of primordial hydrogen, and much smaller amounts of deuterium and lithium. Most of the rest is ordinary hydrogen. (Heavier elements, in tiny, tiny amounts, are created in stars, or by the heat and pressure of exploding stars, and came much later.) These abundances of deuterium and helium are what astronomers observe, which is considered strong evidence for the big bang theory.

The fourth pillar is the (accidental) 1965 discovery by Arno Penzias and Robert Wilson of the cosmic microwave background (CMB) left over from the big bang. The pair of Bell Lab engineers had been trying to nail down a source of noise in microwave reception from the first generation of communications satellites. They saw this remnant radiation everywhere they looked in the sky, and it seemed to have the same strength in every direction. Eventually, after eliminating man-made noise sources, they concluded it was natural, which astronomers confirmed for them.

The uniformity of the CMB is remarkable. The differences represent only sixty parts per million. By comparison, this is about the difference in loudness between a normal conversation and a faint whisper. These cosmic "whispers" contain vital clues to the early days of the universe, as will be explored later in this anthology. The uniformity itself led some to propose inflation, a big bang refinement where, in its earliest times, the universe grew exponentially, so that the potentially large roughness that might have been visible in the CMB is smoothed by intense stretching.

These four pillars—general relativity, the universe's expansion, abundances of light elements, and the cosmic microwave background—support what had become a widely accepted view of how the universe evolved. Some astronomers also had the entirely reasonable expectation that the acceleration Hubble detected was slowing down, being braked by the gravitational pull of all the mass in the universe.

But that is not what they saw. In fact, the acceleration had picked up its pace. And that is the revolution.

It is a revolution where scientists have made "philosophically significant measurements" that challenge current theories, as Saul Perlmutter writes in his article "Supernovae, Dark Energy, and the Accelerating Universe."

For thousands of years, Perlmutter says, "cosmology has been a theorist's domain, where elegant theory was only occasionally endangered by inconvenient facts." But we are now living "in an unusual time, perhaps the first golden age of empirical cosmology," when theory must play catch-up with the measurements.

According to Perlmutter, these measurements have revealed that "not only is the universe accelerating, but it apparently consists primarily of mysterious substances" that have been labeled "dark matter" and "dark energy" to further increase our bafflement. The very terms conjure visions of science fiction and fantasy, but the reality described in these articles is perhaps stranger than can be imagined.

Small wonder, then, that some of the authors collected here betray an edge of excitement in their words as they summarize what they've learned so far and what they hope to learn from measurements that will be made in the years to come.

The majority of this anthology consists of articles on the newly discovered acceleration and how it has altered our view of the universe, most immediately by implying the existence of dark energy. The first two chapters are devoted to this theme.

Chapter 3 focuses on dark matter, which is known only through its gravitational effects. The final chapter explores topics closer to home—the evolution of the chemical elements that make life possible and the creation of solar systems.

This collection of articles was assembled to give an overview of astronomy's cutting edge, where the subatomic is just as important as the galactic for unraveling the unknown in the universe. Think of these articles as dispatches from explorers of a newly discovered realm where a mysterious dark energy is tearing space and time apart. —RA

The Dawning of a New Age

One of the greatest surprises to emerge in the new view of cosmology is the close connection between the enormously large and the incredibly small. This was detailed when in 1999 the National Research Council's Board on Physics and Astronomy made a presentation showing that research in the two disciplines were headed in the same direction. In attendance at that presentation was NASA administrator Daniel Goldin. He was so inspired by what he saw, that he asked the board to assess the science opportunities in the intersection of physics and astronomy, and to devise a plan to make them happen.

The result was the book Connecting Quarks with the Cosmos: Eleven Science Questions for the New Century in 2003. "More than ever before," the scientists wrote, "astronomical discoveries are driving the frontiers of elementary particle physics, and more than ever before our knowledge of the *elementary particles is driving progress in understanding the universe and its contents."*

The following is the executive summary of the content covered more extensively in the book. Although this summary is intended for members of the U.S. government who have the power to increase the funding for the research in physics and astronomy, it provides a great overview of the most important knowledge gaps that, when filled, will mean one of the greatest leaps for science in history. Some of these questions may be answered within the next ten years. For example, physicists are just now learning how to capture and study neutrinos in order to answer the fifth question! —RA

"Executive Summary"

Connecting Quarks with the Cosmos: Eleven Science Questions for the New Century
by the Board on Physics and Astronomy, 2003

We are at a special moment in our journey to understand the universe and the physical laws that govern it. More than ever before astronomical discoveries are driving the frontiers of elementary particle physics, and more than ever before our knowledge of the elementary particles is driving progress in understanding the universe and its contents. The Committee on the Physics of the Universe was convened in recognition of the deep connections that exist between quarks and the cosmos.

The Questions

Both disciplines—physics and astronomy—have seen stunning progress within their own realms of study in the past two decades. The advances made by physicists in understanding the deepest inner workings of matter, space, and time and by astronomers in understanding the universe as a whole as well as the objects within it have brought these scientists together in new ways. The questions now being asked about the universe at its two extremes—the very large and the very small—are inextricably intertwined, both in the asking and in the answering, and astronomers and physicists have been brought together to address questions that capture everyone's imagination.

The answers to these questions strain the limits of human ingenuity, but the questions themselves are crystalline in their clarity and simplicity. In framing this report, the committee has seized on 11 particularly direct questions that encapsulate most of the physics and astrophysics discussed here. They do not cover all of these fields but focus instead on the interface between them. They are also questions that we have a good chance of answering in the next decade, or should be thinking about answering in following decades. Among them are the most profound questions that human beings have ever posed about the cosmos. The fact that they are ripe now, or soon will be, further highlights how exciting the possibilities of this moment are. The 11 questions are these:

What Is Dark Matter?

Astronomers have shown that the objects in the universe, from galaxies a million times smaller than ours to the largest clusters of galaxies, are held together by a form of matter different from what we are made of and that gives off no light. This matter probably consists of one or more as-yet-undiscovered elementary particles, and aggregations of it produce the gravitational pull leading to the formation of galaxies and large-scale structures in the universe. At the same time these particles may be streaming through our Earth-bound laboratories.

What Is the Nature of Dark Energy?

Recent measurements indicate that the expansion of the universe is speeding up rather than slowing down. This discovery contradicts the fundamental idea that gravity is always attractive. It calls for the presence of a form of energy, dubbed "dark energy," whose gravity is repulsive and whose nature determines the destiny of our universe.

How Did the Universe Begin?

There is evidence that during its earliest moments the universe underwent a tremendous burst of expansion, known as inflation, so that the largest objects in the universe had their origins in subatomic quantum fuzz. The underlying physical cause of this inflation is a mystery.

Did Einstein Have the Last Word on Gravity?

Black holes are ubiquitous in the universe, and their intense gravity can be explored. The effects of strong gravity in the early universe have observable consequences. Einstein's theory should work as well in these situations as it does in the solar system. A complete theory of gravity should incorporate quantum effects—Einstein's theory of gravity does not—or explain why they are not relevant.

What Are the Masses of the Neutrinos, and How Have They Shaped the Evolution of the Universe?

Cosmology tells us that neutrinos must be abundantly present in the universe today. Physicists have found evidence that they have a small mass, which implies that cosmic neutrinos account for as much mass as do stars. The pattern of neutrino masses can reveal much about how nature's forces are unified, how the elements in the periodic table were made, and possibly even the origin of ordinary matter.

How Do Cosmic Accelerators Work and What Are They Accelerating?

Physicists have detected an amazing variety of energetic phenomena in the universe, including beams of particles of unexpectedly high energy but of unknown origin. In laboratory accelerators, we can produce beams of energetic particles, but the energy of these cosmic beams far exceeds any energies produced on Earth.

Are Protons Unstable?

The matter of which we are made is the tiny residue of the annihilation of matter and antimatter that emerged from the earliest universe in not-quite-equal amounts. The existence of this tiny imbalance may be tied to a hypothesized instability of protons, the simplest form of matter, and to a slight preference for the formation of matter over antimatter built into the laws of physics.

What Are the New States of Matter at Exceedingly High Density and Temperature?

The theory of how protons and neutrons form the atomic nuclei of the chemical elements is well developed. At higher densities, neutrons and protons may dissolve into an undifferentiated soup of quarks and gluons, which can be probed in heavy-ion accelerators. Densities beyond nuclear densities occur and can be probed in neutron stars, and still higher densities and temperatures existed in the early universe.

Are There Additional Space-Time Dimensions?

In trying to extend Einstein's theory and to understand the quantum nature of gravity, particle physicists have posited the existence of space-time dimensions beyond those that we know. Their existence could have implications for the birth and evolution of the universe, could affect the interactions of the fundamental particles, and could alter the force of gravity at short distances.

How Were the Elements from Iron to Uranium Made?

Scientists' understanding of the production of elements up to iron in stars and supernovae is fairly complete. Important details concerning the production of the elements from iron to uranium remain puzzling.

Is a New Theory of Matter and Light Needed at the Highest Energies?

Matter and radiation in the laboratory appear to be extraordinarily well described by the laws of quantum mechanics, electromagnetism, and their unification as quantum electrodynamics. The universe presents us with places and objects, such as neutron stars and the sources of gamma ray bursts, where the conditions are far more extreme than anything we can reproduce on Earth that can be used to test these basic theories.

Each question reveals the interdependence between discovering the physical laws that govern the universe and understanding its birth and evolution and the objects within it. The whole of each question is greater than the sum of the astronomy part and the physics part of which it is made. Viewed from a perspective that includes both astronomy and physics, these questions take on a greater urgency and importance.

Taken as a whole, the questions address an emerging model of the universe that connects physics at the most microscopic scales to the properties of the universe and its contents on the largest physical scales. This bold construction relies on extrapolating physics tested today in the laboratory and within the solar system to the most exotic astronomical objects and to the first moments of the universe. Is this ambitious extrapolation correct? Do we have a coherent model? Is it consistent? By measuring the basic properties of the universe, of black holes, and of elementary particles in very different ways, we can either falsify this ambitious vision of the universe or establish it as a central part of our scientific view.

The science, remarkable in its richness, cuts across the traditional boundaries of astronomy and physics. It brings together the frontier in the quest for an understanding of the very nature of space and time with the frontier in the quest for an understanding of the origin and earliest evolution of the universe and of the most exotic objects within it.

Realizing the extraordinary opportunities at hand will require a new, crosscutting approach that goes beyond viewing this science as astronomy or physics and that brings to bear the techniques of both astronomy and physics, telescopes and accelerators, and ground- and space-based instruments. The goal then is to create a new strategy. The obstacles are sometimes disciplinary and sometimes institutional, because the science lies at the interface of two mature disciplines and crosses the boundaries of three U.S. funding agencies: the Department of Energy (DOE), the National Aeronautics and Space Administration (NASA), and the National Science Foundation (NSF). If a cross-disciplinary, cross-agency approach can be mounted, the committee believes that a great leap can be made in understanding the universe and the laws that govern it.

The second part of the charge to the committee was to recommend a plan of action for NASA, NSF, and DOE. In Chapter 7, it does so. First, the committee reviewed the projects in both astronomy and physics that have been started (or are slated to start) and are especially relevant to realizing the science opportunities that have been identified. Next, it turned its attention to new initiatives that will help to answer the 11 questions. The committee summarizes its strategy in the seven recommendations described below.

Within these recommendations the committee discusses six future projects that are critical to realizing the great opportunities before us. Three of them—the Large Synoptic Survey Telescope, the Laser Interferometer Space Antenna, and the Constellation-X Observatory—were previously identified and recommended for priority by the 2001 National Research Council decadal survey of astronomy,

Astronomy and Astrophysics in the New Millennium, on the basis of their ability to address important problems in astronomy. The committee adds its support, on the basis of the ability of the projects to also address science at the intersection of astronomy and physics. The other three projects—a wide-field telescope in space; a deep underground laboratory; and a cosmic microwave background polarization experiment—are truly new initiatives that have not been previously recommended by other NRC reports. The committee hopes that these new projects will be carried out or at least started on the same time scale as the projects discussed in the astronomy decadal survey, i.e., over the next 10 years or so.

The initiative outlined by the committee's recommendations can realize many of the special scientific opportunities for advancing our understanding of the universe and the laws that govern it, but not within the budgets of the three agencies as they stand. The answer is not simply to trim the existing programs in physics and astronomy to make room for these new projects, because many of these existing programs—created to address exciting and timely questions squarely within physics or astronomy—are also critical to answering the 11 questions at the interface of the two disciplines. New funds will be needed to realize the grand opportunities before us. These opportunities are so compelling that some projects have already attracted international partners and others are likely to do so.

The Recommendations

Listed below are the committee's seven recommendations for research and research coordination needed to address the 11 science questions.

Measure the polarization of the cosmic microwave background with the goal of detecting the signature of inflation.

The committee recommends that NASA, NSF, and DOE undertake research and development to bring the needed experiments to fruition.

Cosmic inflation holds that all the structures we see in the universe today—galaxies, clusters of galaxies, voids, and the great walls of galaxies—originated from subatomic quantum fluctuations that were stretched to astrophysical size during a tremendous spurt of expansion (inflation). Quantum fluctuations in the fabric of space-time itself lead to a cosmic sea of gravitational waves that can be detected by their polarization signature in the cosmic microwave background radiation.

Determine the properties of dark energy.

The committee supports the Large Synoptic Survey Telescope project, which has significant promise for shedding light on the dark energy. The committee further recommends that NASA and DOE work together to construct a wide-field telescope in space to determine the expansion history of the universe and fully probe the nature of dark energy.

The discovery that the expansion of the universe is speeding up and not slowing down through the study of distant supernovae has revealed the presence of a mysterious new energy form that accounts for two-thirds of all the matter and energy in the universe. Because of its diffuse nature, this energy can only be probed through its effect on the expansion of the universe. The NRC's most recent astronomy decadal survey recommended building the Large Synoptic Survey Telescope to study transient phenomena in the universe; the telescope will also have significant ability to probe dark energy. To fully characterize the expansion history and probe the dark energy will require a wide-field telescope in space (such as the Supernova/Acceleration Probe) to discover and precisely measure the light from very distant supernovae.

Determine the neutrino masses, the constituents of the dark matter, and the lifetime of the proton.

The committee recommends that DOE and NSF work together to plan for and to fund a new generation of experiments to achieve these goals. It further recommends that an underground laboratory with sufficient infrastructure and depth be built to house and operate the needed experiments.

Neutrino mass, new stable forms of matter, and the instability of the proton are all predictions of theories that unify the forces of nature. Fully addressing all three questions requires a laboratory that is well shielded from the cosmic-ray particles that constantly bombard the surface of Earth.

Use space to probe the basic laws of physics.

The committee supports the Constellation-X and Laser Interferometer Space Antenna missions, which hold great promise for studying black holes and for testing Einstein's theory in new regimes. The committee further recommends that the agencies proceed with an advanced technology program to develop instruments capable of detecting gravitational waves from the early universe.

The universe provides a laboratory for exploring the laws of physics in regimes that are beyond the reach of terrestrial laboratories. The NRC's most recent astronomy decadal survey recommended the Constellation-X Observatory and the Laser Interferometer Space Antenna on the basis of their great potential for astronomical discovery. These missions will be able to uniquely test Einstein's theory in regimes where gravity is very strong: near the event horizons of black holes and near the surfaces of neutron stars. For this reason, the committee adds its support for the recommendations of the astronomy decadal survey.

Determine the origin of the highest-energy gamma rays, neutrinos, and cosmic rays.

The committee supports the broad approach already in place and recommends that the United States ensure the timely completion and operation of the Southern Auger array.

The highest-energy particles accessible to us are produced by natural accelerators throughout the universe and arrive on Earth as high-energy gamma rays, neutrinos, and cosmic rays. A full understanding of how these particles are produced and accelerated could shed light on the unification of nature's forces. The Southern Auger array in Argentina is crucial to solving the mystery of the highest-energy cosmic rays.

Discern the physical principles that govern extreme astrophysical environments through the laboratory study of high-energy-density physics.

The committee recommends that the agencies cooperate in bringing together the different scientific communities that can foster this rapidly developing field.

Unique laboratory facilities such as high-power lasers, high-energy accelerators, and plasma confinement devices can be used to explore physics in extreme environments as well as to simulate the conditions needed to understand some of the most interesting objects in the universe, including gamma-ray bursts. The field of high-energy-density physics is in its infancy, and to fulfill its potential, it must draw on expertise from astrophysics, laser physics, magnetic confinement and particle beam research, numerical simulation, and atomic physics.

Realize the scientific opportunities at the intersection of physics and astronomy.

The committee recommends establishment of an interagency initiative on the physics of the universe, with the participation of DOE, NASA, and NSF. This initiative should provide structures for joint planning and mechanisms for joint implementation of cross-agency projects.

The scientific opportunities the committee identified cut across the disciplines of physics and astronomy as well as the boundaries of DOE, NASA, and NSF. No agency has complete ownership of the science. The unique capabilities of all three, as well as cooperation and coordination between them, will be required to realize these special opportunities.

The Committee on the Physics of the Universe believes that recent discoveries and technological developments make the time ripe to greatly advance our understanding of the origin and fate of the universe and of the laws that govern it. Its 11 questions convey the magnitude of the opportunity before us. The committee believes that implementing these seven recommendations will greatly advance our understanding of the universe and perhaps even our place within it.

Reprinted with permission from (Connecting Quarks with the Cosmos: Eleven Science Questions for the New Century) © (2003) by the National Academy of Sciences, courtesy of the National Academies Press, Washington, D.C.

The universe is expanding. As a result, "galaxies are being swept apart from each other at colossal speeds" as space

expands, author Wendy Freedman writes in the next article.

This expansion was first discovered in 1929 by Edwin Hubble, who also determined that the farther away from our Milky Way a galaxy is, the faster it moves away from us. So, the expansion rate, named the Hubble constant, not only measures the speed of a galaxy—kilometers per second—but it also tells us how the speed increases based on its distance away from us, typically measured in megaparsecs, where one megaparsec is 3.26 million light-years.

In principle, all we need to estimate Hubble's constant is a galaxy's speed and distance from us. But as Freedman discusses, this is much trickier than it seems, although the effort is worthwhile because the value tells us a great deal about the universe's age, composition, evolution, and fate.

Freedman describes the international effort she led (the Hubble Key Project) to nail down the value—72 km/s/megaparsec—using measurements taken with the Hubble Space Telescope. —RA

"On the Age of the Universe"
by Wendy L. Freedman
Daedalus, Winter 2003

How did the world begin? How old is it? Do mysterious and invisible forces determine its fate?

Surprisingly enough, such questions are now at the forefront of scientific research.

Over the past century, old ideas about the cosmos and our place in it have been dramatically overturned. We now know that the Sun does not occupy the center of the universe, and that in addition to our own Milky Way, space is filled with hundreds of billions of other galaxies. Even more astonishingly, we know that the universe itself is expanding everywhere, and that as space expands, galaxies are being swept apart from each other at colossal speeds.

In the last few years, tantalizing hints have begun to appear that the expansion of the universe is even accelerating. These results imply the existence of a mysterious force able to counter the attraction of gravity. The origin and nature of this force currently defy explanation. But astronomers have reason to hope that ongoing research will soon resolve some of the deepest riddles of nature.

It was Edwin Hubble, a Carnegie Astronomer based in Pasadena, California, who first learned that the universe was expanding; in 1929, he discovered that the farther away from our Milky Way galaxies are, the faster they are moving apart. A few years before, Albert Einstein in his general theory of relativity had published a mathematical formula for the evolution of the universe. Einstein's equations, like Hubble's observations, implied that the universe must once have been much denser and hotter. These results suggested that the universe began with an intense explosion, a "big bang."

The big bang model has produced a number of testable predictions. For example, as the universe expands, the hot radiation produced by the big bang will cool and pervade the universe—thus we should see heat in every direction we look. Big bang theory predicts that by today the remnant radiation should have cooled to a temperature of only 3 degrees above absolute zero (corresponding to a temperature of -270 degrees Celsius). Remarkably, this radiation has been detected. In 1965, two radio astronomers, Arnold Penzias and Robert Wilson, discovered this relic radiation during a routine test of communications dishes, a discovery for which they were awarded the Nobel Prize.

The current expansion rate of the universe, known as the Hubble constant, determines the size of the observable universe and provides constraints on competing models of the evolution of the universe. For decades, an uncertainty of a factor of two in measurements of the Hubble constant existed. (Indeed, determining an accurate value for the Hubble constant was one of the main reasons for building the Hubble Space Telescope.) However, rapid progress has been made recently in resolving the differences. New, sensitive instruments on telescopes, some flying aboard the Hubble Space Telescope, have led to great strides in the measurement of distances to galaxies beyond our own.

In theory, determining the Hubble constant is simple: one need only measure distance and velocity. But in practice, making such measurements is difficult.

It is hard to devise a means to measure distances over cosmological scales accurately. And measuring velocity is complicated by the fact that neighboring galaxies tend to interact gravitationally, thereby perturbing their motions. Uncertainties in distances and in velocities then lead to uncertainties in their ratio, the Hubble constant.

Velocities of galaxies can be calculated from the observed shift of lines (due to the presence of chemical elements such as hydrogen, iron, oxygen) in the spectra of galaxies. There is a familiar analogous phenomenon for sound known as the Doppler effect, which explains, for instance, why the pitch of an oncoming train changes as the train approaches and then recedes from us. As galaxies move away from us, their light is similarly shifted and stretched to longer (redder) wavelengths, a phenomenon referred to as redshift. This shift in wavelength is proportional to velocity.

Measuring distances presents a greater challenge, which has taken the better part of a century to resolve. Most distances in astronomy cannot be measured directly because the size scales are simply too vast. For the very nearest stars, distances can be measured using a method called parallax. This uses the baseline of the Earth's orbit, permitting the distance to be calculated using simple, high-school trigonometry. However, this technique currently can be applied reliably only for relatively nearby stars within our own galaxy.

In order to measure the distance of more remote stars and galaxies, astronomers identify objects that exhibit a constant, known brightness, or a brightness that is related to another measurable quantity. The distance is then calculated using the inverse square law of radiation, which states that the apparent brightness of an object falls off in proportion to the square of its distance from us. The effects of the inverse square law are easy to see in everyday life—say if we compare the faint light of a train in the distance with the brilliant light as the train bears down close to us.

To get a sense of the (astronomical) scales we are talking about, the nearest star to us is about 4 light-years away. One light-year is the distance that light can travel within a year moving at the enormous speed of 186,000 miles per second. At this speed, light circles the Earth more than 7 times in 1 second. For comparison, the "nearby" Andromeda galaxy lies at a distance of about 2 million light-years. And the most distant galaxies visible to us currently are about 13 billion light-years away. That is to say, the light that left them 13 billion years ago is just now reaching us, and we are seeing them as they were 13 billion years ago, long before the Sun and Earth had even formed (4.6 billion years ago).

Until recently, one of the greatest challenges to measuring accurate distances was a complication caused by the presence of dust grains manufactured by stars and scattered throughout interstellar space. This dust, located in the regions between stars, absorbs and scatters light. If no correction is made for its effects, objects appear fainter and therefore apparently, but erroneously, farther away than they actually are. Fortunately, dust makes objects appear not only fainter, but also redder. By making measurements at more than one wavelength, this color dependence provides a powerful means of correcting for the presence of dust and allowing correct distances to be derived.

Currently, the most precise method for measuring distances is based on the observations of stars named Cepheid variables. The atmospheres of these stars pulsate in a very regular cycle, on timescales ranging from 2 days to a few months. The brighter the Cepheid, the more slowly it pulsates, a property discovered by astronomer Henrietta Leavitt in 1908. This unique relation allows the distance to be obtained, again using the inverse square law of radiation—that is, it allows the intrinsic brightness of the Cepheid to be predicted from its observed period, and its distance from Earth to be calculated from its observed, apparent brightness.

High resolution is vital for discovering Cepheids in other galaxies. In other words, a telescope must have sufficient resolving power to distinguish individual Cepheids from all the other stars in the galaxy. The resolution of the Hubble Space Telescope is about ten times better than can be generally obtained through Earth's turbulent atmosphere. Therefore galaxies within a volume about a thousand times greater than accessible to telescopes from Earth could be measured for the first time with Hubble. With it,

distances to galaxies with Cepheids can be measured
relatively simply out to the nearest massive clusters
of galaxies some 50 to 70 million light-years away.
(For comparison, the light from these galaxies began
its journey about the time of the extinction of the
dinosaurs on Earth.)

Beyond this distance, other methods—for example,
bright supernovae or the luminosities of entire galaxies—
are employed to extend the extragalactic distance scale
and measure the Hubble constant. Supernovae are cat-
aclysmic explosions of stars near the end of their lives.
The intrinsic luminosities of these objects are so great
that for brief periods, they may shine as bright as an
entire galaxy. Hence, they may be seen to enormous
distances, as they have been discerned out to about half
the radius of the observable universe. Unfortunately,
for any given method of measuring distances, there
may be uncertainties that are as yet unknown.
However, by comparing several independent methods,
a limit to the overall uncertainty of the Hubble constant can be obtained. This was one of the main aims
of the Hubble Key Project.

This project was designed to use the excellent resolving power of the Hubble Space Telescope to discover and
measure Cepheid distances to galaxies, and to determine
the Hubble constant by applying the Cepheid calibration
to several methods for measuring distances further out in
the Hubble expansion. The Key Project was carried out by
a group of about 30 astronomers, and the results were
published in 2001. Distances measured using Cepheids

were used to set the absolute distance scale for 5 different methods of measuring relative distances. The combined results yield a value of the Hubble constant of 72 (in units of kilometers per second per megaparsec, where 1 megaparsec corresponds to a distance of 3.26 million light-years), with an uncertainty of 10 percent. (The previous range of these measurements was 40 to 100 in these units.) Unlike the situation earlier, all of the different methods yield results in good agreement to within their respective measurement uncertainties.

The Hubble constant is the most important parameter in gauging the age of the universe. However, in order to determine a precise age, it is important to know how the current expansion rate differs from past rates. If the universe has slowed down or speeded up over time, then the total length of time over which it has been expanding will differ accordingly. Is the universe slowing down (as expected if the force of gravity has been retarding its expansion)? If so, the expansion would have been faster in the past before the effects of gravity slowed it down, and the age estimated for the universe would be younger than if it had always been expanding at a constant rate.

Indeed, this deceleration is what astronomers expected to find as they looked further back in time. The calculation for a Hubble constant of 72 and a universe with a slowing expansion rate yields an age for the universe of about 9 billion years. This would be fine, except for one not-so-small detail from other considerations: the measured ages of stars.

The best estimates of the oldest stars in the universe are obtained from studying globular clusters, systems of stars that formed early in the history of our galaxy. Stars spend most of their lifetimes undergoing the nuclear burning of hydrogen into helium in their central cores. Detailed computer models of the evolution of such stars compared with observations of them in globular clusters suggest they are about 12 or 13 billion years old—apparently older than the universe itself. Obviously, this is not possible.

The resolution of this paradox appears to rest in a newly discovered property of the universe itself. A wealth of new data over the past few years has begun to evolutionize cosmology. Probably the most surprising result is the increasing evidence that instead of decelerating as expected, the universe is accelerating! One implication is the existence of a form of energy that is repulsive, acting against the inward pull of gravity. Astronomers refer to this newly discovered universal property of the universe as "dark energy."

Before the expansion of the universe was discovered, Einstein's original mathematical equation describing the evolution of the universe in general relativity contained a term that he called the cosmological constant. He introduced this term to prevent any expansion (or contraction) of the universe, as it was thought that the universe was static. After Hubble discovered the expansion, Einstein referred to the cosmological constant as his greatest blunder. He had missed the opportunity to predict the expansion.

However, a recent discovery suggests that, although the universe is expanding, the term in Einstein's equation may have been correct after all: it may represent the dark energy. In a universe with a Hubble constant of about 70, and with matter contributing one-third and dark energy providing approximately two-thirds of the overall mass plus energy density, the resulting estimated age for the universe is 13 billion years, in very good agreement with the ages derived from globular clusters.

It is too soon yet to know whether the existence of dark energy will be confirmed with future experiments. But to the surprise of an initially skeptical community of astronomers and physicists, several independent observations and experiments are consistent with this theory. Perhaps most exciting is the prospect of learning more about an entirely new form of mysterious energy, a property of the universe that to date has evaded all explanation.

The dark energy observed is smaller by at least 10 billion, billion, billion, billion, billion, billion times than the best theories of elementary particle physics would predict from first principles. Hence, by studying the behavior of the universe, astronomers are posing new challenges to fundamental physics. It is often the case in science that as old questions are resolved, novel, perhaps even more exciting, questions are uncovered. The next decade promises to be a fruitful one in addressing profound questions about the nature of the universe we live in.

Wendy Freedman, "On the Age of the Universe," *Daedalus*, 132:1 (Winter, 2003), pp. 122–125. © 2003 by the Massachusetts Institute of Technology and the American Academy of Arts and Sciences. Reprinted with permission.

The next article describes the momentous year of 1998, when two studies of old, distant exploding stars "overturned the prevailing belief that the cosmos has been slowing down its rate of expansion ever since the Big Bang," as author Ron Cowen writes. Instead, Cowen explores how the rate has sped up and the universe is "flying apart faster than ever before."

This result was entirely unexpected, but it appeared to "unify elements of a cosmic portrait" that had emerged in the preceding decade, and stitched together such "disparate concepts as energy associated with empty space, invisible matter in the universe, and the curvature of the cosmos."

The backdrop for Cowen's article is The Nature of the Universe debate, the third Great Debate in Astronomy in a series held in the late 1990s that had been inspired by the first one in 1920. The latter-day debates were organized by astronomers to highlight major quests in astronomy and astrophysics at the turn of the millennium. All four debates were held at the Smithsonian Institution and open to the public. The 1920 debate and the first pair of debates in the 1990s addressed the size of the universe. But the final debate in 1998 crowned *them all, with its sweeping inquiry into the very nature of the universe made possible by the discoveries of that year. So great was this advance of knowledge that one of the 1998 debaters can be forgiven for posing the question, "Is cosmology solved?" —RA*

"The Greatest Story Ever Told"
by Ron Cowen
Science News, **December 19 & 26, 1998**

Is Cosmology Solved?

Scientists don't often make great debaters. Rather than dealing with absolutes in black and white, they tend to invoke qualifiers and caveats in shades of gray. But cosmologist Michael S. Turner, whose hand-drawn view-graphs are so colorful that they have adorned the walls of an art gallery, isn't the typical scientist, and 1998 hasn't been the typical year for the study of the universe.

In an October forum—billed as "The Nature of the Universe Debate: Cosmology Solved?"—Turner, who is at the University of Chicago and the Fermi National Accelerator Laboratory in Batavia, Ill., put forth an assertion as bold as his drawings: For the first time in history, cosmologists have developed a consistent framework that integrates the origin, evolution, and current appearance of the universe.

Turner's opponent in the debate, Jim Peebles of Princeton University, took a more conservative view of recent progress in deciphering the cosmos. He prescribed caution in concluding that the key pieces of the cosmic puzzle have all been revealed.

The past year could mark a turning point for cosmology, Turner told a packed auditorium at the Smithsonian Institution's National Museum of Natural History in Washington, D.C.

Clearly, 1998 began with a jolt: Two rival teams studying the titanic explosion of distant, elderly stars overturned the prevailing belief that the cosmos has been slowing down its rate of expansion ever since the Big Bang. In fact, they reported, the universe is actually flying apart faster than ever before (SN: 3/21/98, p. 185; 10/31/98, p. 277).

Although entirely unexpected, that recent finding and others appear to unify elements of a cosmic portrait that have emerged over the past decade, Turner says. Stitching together such disparate concepts as energy associated with empty space, invisible matter in the universe, and the curvature of the cosmos, the new reports may turn out to mark a watershed for cosmology. Their impact could be every bit as important as the discovery more than 3 decades ago of the whisper of radiation left over from the Big Bang.

In 1964, two physicists at Bell Laboratories in Holmdel, N.J., stumbled upon key evidence for the Big Bang. Scanning the sky with a radio receiver, they discovered a faint, uniform crackling. The pervasive nature of the signal and its intensity over a range of frequencies indicated that the radiation could not have come from the universe today. Instead, Arno Penzias and Robert Wilson concluded, it represents the radiation produced by the cosmos when it was young and extremely hot.

This radiation, known as the cosmic microwave background, is one of the cornerstones of the Big Bang theory. Along with measurements of the abundance of light elements forged just after the birth of the universe, the microwave background provides evidence that the universe began with the explosive expansion of a dense, hot soup of subatomic particles and radiation.

A fog of electrons pervaded the infant universe. For thousands of years after the Big Bang, radiation did not stream freely into space but was repeatedly absorbed and scattered by these charged particles.

About 300,000 years after the Big Bang, the universe became cool enough for the electrons to combine with nuclei. This lifted the fog, enabling radiation to travel unimpeded. Shifted to longer wavelengths by the expansion of the universe, this relic radiation is today detected as microwaves and far-infrared light. It provides a snapshot of the universe when it was 300,000 years old.

The Big Bang model has been phenomenally successful in explaining the events that took place beginning one-hundredth of a second after the birth of the universe. But by 1980, scientists trying to elucidate even earlier cosmic events were pushing the limits of the theory. The Big Bang model offers no explanation for the explosion itself, notes Turner. The dynamite that produced the Big Bang remains elusive.

The model has other shortcomings. It does not reveal the nature of the matter that fills the universe. Nor does it explain why the young cosmos was so smooth and uniform and how tiny fluctuations in the density of the early universe could give rise to the lumpy collection of galaxies, clusters of galaxies, and superclusters seen today.

A theory known as inflation, developed and refined during the 1980s, provides a partial answer to these riddles. In this theory, the cosmos undergoes an extremely short but prodigious growth spurt. In just 10^{-32} second, the universe expanded more than it has in the 13 billion years or so that has elapsed since (SN: 6/7/97, p. 354).

This growth spurt captured chance subatomic fluctuations in energy and inflated them to macroscopic proportions. The action transformed the fluctuations into regions of slightly higher and lower density. Over time, gravity molded these variations into the spidery network of galaxies and voids seen in the universe today.

In inflationary cosmology, quantum fluctuations provide the energy for the expansion. According to quantum theory, the vacuum of space is far from empty. It seethes with particles and antiparticles constantly being created and destroyed. Energy from this vacuum can be tapped and is more than sufficient to trigger the era of explosive expansion dubbed the Big Bang.

This scenario gained support from a 1992 discovery: The microwave background does not have a uniform temperature but is full of hot spots and cold spots. The variations, a few ten-thousandths of a kelvin, were detected by the Cosmic Background Explorer (COBE) satellite. They are thought to correspond to slight variations in the distribution of matter at the moment when light and matter parted company, and radiation streamed freely into space. This finding was hailed as proof that microscopic lumps in the infant cosmos, no bigger than about 10^{-23} centimeters across, were the seeds for the galaxies and other large-scale structures we see today.

Inflation also explains the overall uniformity of the universe. Conventional Big Bang cosmology cannot account for how regions of the universe separated by distances so large that they have never even exchanged light signals can look so similar to each other. According to inflation theory, the universe began with regions so tiny that they were homogeneous. These regions then expanded into volumes vastly bigger than astronomers can ever observe.

Inflation makes the cosmos not only uniform but also flat. Any curvature to space-time is stretched out by the expansion, like a cosmic version of a balloon stretched to enormous proportions.

Over the past 4 years, nearly 20 ground-based and balloon-borne telescopes began measuring variations in the temperature of the cosmic microwave background over small spatial scales. The pattern of variations is known to be sensitive to the shape of the cosmos (SN: 2/21/98, p. 123). The measurements are not yet conclusive, but they suggest that the geometry of the universe is indeed as flat as the inflation theory would predict.

For the universe to be flat, astronomers calculate that it must contain a critical density of material. A variety of observations, however, including measurements taken over the past year, has revealed that the universe comes up short: It doesn't have nearly enough matter to be flat.

Of the four types of lightweight nuclei forged in the Big Bang, deuterium is the most sensitive indicator of the density of ordinary matter, which is made of protons, neutrons, and electrons. The greater the density of deuterium, the lower the density of ordinary matter (SN: 5/18/96, p. 309).

Because stars burn deuterium, the amount present today is not a good indicator of the primordial abundance, Turner notes. By measuring deuterium in very distant, essentially starless, hydrogen clouds, which hail from a time when the universe was very young, David R. Tytler of the University of California, San Diego and Scott Burles of the University of Chicago this year pinned down the amount of deuterium made in the Big Bang. Their measurements indicate that the density of ordinary matter contributes only 5 percent of the density needed for the universe to be flat.

Astronomers set their sights on clusters of galaxies to estimate the total density of matter in the cosmos, including exotic kinds of matter that would reveal their presence only through their gravitational influence. To weigh these behemoth clusters, the scientists use several methods. They measure the temperature of the X rays the clusters emit, and they determine the random motion of galaxies within a cluster.

With these methods, researchers recently found that the total density of matter is about 40 percent of the critical density.

This result has two profound implications, Turner notes. First, it suggests that most of the matter in the universe is not the familiar stuff that rocks and people are made of. Rather, it's some unseen, exotic material.

This dark matter could be remnants from the earliest, fiery moments of the universe, when high temperatures would have set the stage for the creation of a vast zoo of elementary particles. Slow-moving particles, generically known as cold dark matter, are the best candidates for this exotic material, Turner says. These particles would allow for the pattern of structures seen in the universe today, which indicates that it evolved from the bottom up, forming galaxies, then clusters of galaxies, and so on. Other theories had suggested that the large structures formed first, then fragmented.

The other consequence of the new measurements of matter density is even more startling. If the universe is flat, then there must be something else—a special form of matter or energy (the two are equivalent according to Einstein) that makes up the missing 60 percent of the critical density. Turner dubs this component "funny energy."

This special energy resists the gravitational pull of galaxies, so it distributes itself uniformly throughout the cosmos.

Funny energy "leads to a striking prediction," notes Turner. "The expansion of the universe should be speeding up, rather than slowing down."

How so? According to Einstein's theory of general relativity, gravity derives both from energy and matter and from pressure. The funny energy manifests itself as a negative pressure. If the universe contains a large enough component of funny energy, the net effect of gravity is to exert a repulsive, rather than an attractive, force. The expansion of the universe then accelerates rather than slows down.

In 1998, this bizarre state of affairs received tentative confirmation. Two teams of scientists, including researchers at the University of California, Berkeley and Lawrence Berkeley (Calif.) National Laboratory, examined a distinct type of exploding star, or supernova. Previous studies have suggested that this type, known as a supernova Ia, has the same intrinsic luminosity in both nearby and distant galaxies.

Because the light from a faraway galaxy takes several billion years to reach Earth, telescopes see such a galaxy as it appeared when the universe was younger. If cosmic expansion had recently slowed, then there should be less distance between Earth and a faraway galaxy—and a shorter travel time for light—than if the expansion had maintained its speed. A supernova in a distant galaxy would look slightly brighter in the case of the slowed expansion.

The researchers this year found exactly the opposite. Distant supernovas looked dimmer than expected, indicating that the universe has increased its rate of expansion.

Measurements of the geometry of the cosmos and the gravity within it finally add up, says Turner. "For the very first time, we have a complete and plausible accounting of matter and energy in the universe."

The avalanche of data now and expected over the next few years will go a long way toward explaining the basic features of the universe with a theory rooted in fundamental physics, Turner concludes. "What I want to argue is that in 1998, we had the first key evidence for a theory that takes us well beyond the hot Big Bang cosmology."

Turner's debate opponent, Peebles, argues that nothing is settled until the proverbial fat lady sings—and as far as Peebles is concerned, she hasn't sung yet. Cautioning his colleagues not to go overboard in their enthusiasm, Peebles recalled the words of a cosmologist of an earlier era, Willem de Sitter, who admonished in 1931 that "it should not be forgotten that all this talk about the universe involves a tremendous extrapolation, which is a very dangerous operation."

"Observational advances since then have greatly reduced the danger," says Peebles, "but I think [they] should leave us with a sense of wonder at the successes in probing the large-scale nature of the physical universe and caution in deciding just how well we understand the situation."

"The basic tenets of inflation plus cold dark matter have not yet been confirmed definitively," Turner admits. He contends, however, that a survey under way to map the location of 1 million nearby galaxies and the planned launch of two NASA missions to record the cosmic microwave background in unprecedented detail "could make the case soon."

Peebles raises another criticism, which Turner acknowledges: The existence of an accelerating universe implies that we live during a very special time in the history of the cosmos. This circumstance harks back to the different behavior of the two components— matter and energy—that contribute to the critical density. The amount of mass per unit volume declines as the universe expands, but the energy density, at least in its simplest form, remains the same. Indeed, it is sometimes referred to as the cosmological constant.

Observations suggest that right now, the densities of matter and funny energy are roughly equal. The energy density is just beginning to take over from matter density as the factor controlling cosmic expansion. At an earlier time, when the mass density was higher, the global effect of gravity would have been attractive, and we would not have observed the universe to be accelerating.

The big question, quips Turner, "is the Nancy Kerrigan question: 'Why me, Why now?'" There's only one period when matter and energy densities are comparable, he notes, "and that's today, and we happen to be around."

Although he finds that seeming coincidence "bothersome," Turner doesn't see it undermining the model of an accelerating universe. He says, "Often in science as you answer one question, a new question is raised."

At the end of the debate between Turner and Peebles, moderator Margaret J. Geller of the Harvard-Smithsonian Center for Astrophysics in Cambridge, Mass., took a straw poll. She asked the listeners if they thought that a century from now, the basic concepts exciting astronomers this year would be cornerstones of understanding. Most of the audience thought the concepts would be substantially different.

Even if today's models endure, "solving cosmology does not mean the end of the study of the universe, nor even the beginning of a less exciting period of inquiry," Turner says in a monograph that accompanied his talk.

"A list of today's puzzles is long enough and challenging enough to occupy astrophysicists for decades: What are the objects that make gamma-ray bursts, and how do they work? How do galaxies form stars and light up the sky? How are stars born? . . . Is there life elsewhere in the cosmos? . . .

"With the flood of data coming, the list will only grow longer and more interesting," Turner concludes. "I am confident that there will be plenty of challenges for next century's astrophysicists."

In a demonstration that advancements made in particle physics have implications for the nature of the universe, a recent earthbound experiment cleared up a nagging disagreement regarding the universe's age. An age that, on the one hand, is determined from cosmic microwave background (CMB) (to be discussed later) and, on the other hand, from standard models of the nuclear reaction in old stars, which is discussed in the following article.

Using particle accelerators to mimic conditions inside stars, researchers found evidence that the oldest known stars found in globular clusters are 700 million years older than previously estimated, bringing their ages to about 14 billion years. This is in fair agreement with the latest estimates of about 13.7 billion years from CMB results.

The age adjustment stems from a nuclear reaction inside old stars that have nearly exhausted the hydrogen they are burning to make helium and release energy. This latecomer process activates an alternative helium-making mechanism that involves collisions between protons and the nuclei of carbon, nitrogen, and oxygen. An experiment performed in an underground accelerator found the actual rate of the reaction was about half the accepted value, which was apparently marred by cosmic rays. Conducting the new experiments underground provided some shielding from the rays. —RA

"New Measurement of Stellar Fusion Makes Old Stars Even Older"
by Kim Krieger
Science, May 28, 2004

A key nuclear reaction inside stars takes significantly longer than standard models assume, European researchers have discovered. The result, which nuclear physicists at the Laboratory for Underground Nuclear Astrophysics (LUNA) in Gran Sasso, Italy, report in a pair of online papers, implies that the most ancient star clusters are at least 700 million years older than previously believed.

"The LUNA experiment is beautiful," says John Bahcall, an astrophysicist at the Institute for Advanced Study in Princeton, New Jersey, praising the group as "magically gifted experimentalists."

The LUNA team used an underground particle accelerator at Gran Sasso to measure the speed of the carbon-nitrogen-oxygen (CNO) cycle, one of the pathways by which stars fuse hydrogen into helium, releasing energy. The cycle determines how long it takes a youthful hydrogen-burning star to turn into a giant helium burner. Astrophysicists can estimate the

age of a star on the cusp of that transition by measuring
its mass and then calculating how long it took to reach
its current state.

The CNO cycle, however, is only as fast as its slow-
est step: a nuclear reaction in which the isotope
nitrogen-14 absorbs a proton from hydrogen and turns
into oxygen-15. Researchers had estimated the rate of
the reaction by shooting protons at nitrogen-14 in particle accelerators. But the measurements were marred
by noise from cosmic rays, and astrophysicists suspected they erred on the speedy side.

In papers scheduled to be published in *Physics
Letters B* and *Astronomy and Astrophysics*, the LUNA
researchers report that the limiting step is indeed only
half as rapid as previously assumed. Working 1400
meters underground to shield their detectors from cosmic radiation, they smashed protons into a nitrogen-14
target and then measured the gamma rays the nitrogen
released as it became oxygen-15. The results push the
age of the oldest stars to almost 14 billion years. That's
close to the figure of 13.7 billion years for the age of the
universe that physicists derived from measurements by
the Wilkinson Microwave Anisotropy Probe (*Science*,
14 February 2003, p. 991), although both still have significant uncertainties, Bahcall says.

The team plans to repeat the experiment at more realistic collision energies, says Carlo Broggini, spokesperson
for the LUNA project. The first set of experiments was
run at energies above 140 kilo-electron volts (KeV),
Broggini says. A new gamma ray detector should allow

researchers to study collisions at close to 25 KeV, the peak energy level at which the reaction occurs in stars.

Reprinted with permission from Krieger, Kim, "New Measurement of Stellar Fusion Makes Old Stars Even Older," SCIENCE 304:1226 (2004). © 2004 AAAS.

Cosmologists generally have the universe's clock set to zero at the moment of creation, the so-called big bang. Although there isn't a standard big bang model, a widely accepted version has the universe springing forth from a special quantum field, expanding faster than the speed of light at accelerating rates in an early period called inflation.

While inflation explains much of what we observe, its singular nature at zero time is troublesome, with its suggestion that everything was once squeezed into zero size.

The approaches for dealing with this bothersome detail are the focus of the next article. Author Gabriele Veneziano describes two prominent scenarios that use the string theory to explain the creation of the universe, radically proposing a ten-dimensional universe populated by vibrating stringlike entities that are the basic building block of matter.

The scenario Veneziano champions, the pre–big bang scenario, assumes the universe *came from an unusually large black hole that had rebounded after it squeezed the universe to the smallest possible size allowed by string theory.*

The ekpyrotic (conflagration) scenario assumes the big bang was the collision of our universe with another, both of them flat and floating about in dimensional space as "D-branes," each attracted to the other. —RA

"The Myth of the Beginning of Time"
by Gabriele Veneziano
Scientific American, May 2004

String theory suggests that the big bang was not the origin of the universe but simply the outcome of a preexisting state

Was the big bang really the beginning of time? Or did the universe exist before then? Such a question seemed almost blasphemous only a decade ago. Most cosmologists insisted that it simply made no sense— that to contemplate a time before the big bang was like asking for directions to a place north of the North Pole. But developments in theoretical physics, especially the rise of string theory, have changed their perspective. The pre-bang universe has become the latest frontier of cosmology.

The new willingness to consider what might have happened before the bang is the latest swing of an intellectual pendulum that has rocked back and forth for millennia. In one form or another, the issue of the ultimate beginning has engaged philosophers and theologians in nearly every culture. It is entwined with a grand set of concerns, one famously encapsulated in an 1897 painting by Paul Gauguin: *D'ou venons-nous? Que sommes-nous? Ou allons-nous?* "Where do we come from? What are we? Where are we going?" The piece depicts the cycle of birth, life and death—origin, identity and destiny for each individual—and these personal concerns connect directly to cosmic ones. We can trace our lineage back through the generations, back through our animal ancestors, to early forms of life and protolife, to the elements synthesized in the primordial universe, to the amorphous energy deposited in space before that. Does our family tree extend forever backward? Or do its roots terminate? Is the cosmos as impermanent as we are?

The ancient Greeks debated the origin of time fiercely. Aristotle, taking the no beginning side, invoked the principle that out of nothing, nothing comes. If the universe could never have gone from nothingness to somethingness, it must always have existed. For this and other reasons, time must stretch eternally into the past and future. Christian theologians tended to take the opposite point of view. Augustine contended that God exists outside of space and time, able to bring these constructs into existence as surely as he could forge other aspects of our world. When asked, "What was God doing *before* he created the world?" Augustine answered, "Time itself being part of God's creation, there was simply no *before*!"

Einstein's general theory of relativity led modern cosmologists to much the same conclusion. The theory holds that space and time are soft, malleable entities. On the largest scales, space is naturally dynamic, expanding or contracting over time, carrying matter like driftwood on the tide. Astronomers confirmed in the 1920s that our universe is currently expanding: distant galaxies move apart from one another. One consequence, as physicists Stephen Hawking and Roger Penrose proved in the 1960s, is that time cannot extend back indefinitely. As you play cosmic history backward in time, the galaxies all come together to a single infinitesimal point, known as a singularity—almost as if they were descending into a black hole. Each galaxy or its precursor is squeezed down to zero size. Quantities such as density, temperature and spacetime curvature become infinite. The singularity is the ultimate cataclysm, beyond which our cosmic ancestry cannot extend.

Strange Coincidence

The unavoidable singularity poses serious problems for cosmologists. In particular, it sits uneasily with the high degree of homogeneity and isotropy that the universe exhibits on large scales. For the cosmos to look broadly the same everywhere, some kind of communication had to pass among distant regions of space, coordinating their properties. But the idea of such communication contradicts the old cosmological paradigm.

To be specific, consider what has happened over the 13.7 billion years since the release of the cosmic microwave background radiation. The distance between galaxies has grown by a factor of about 1,000 (because of the expansion), while the radius of the observable universe has grown by the much larger factor of about 100,000 (because light outpaces the expansion). We see parts of the universe today that we could not have seen 13.7 billion years ago. Indeed, this is the first time in cosmic history that light from the most distant galaxies has reached the Milky Way.

Nevertheless, the properties of the Milky Way are basically the same as those of distant galaxies. It is as though you showed up at a party only to find you were wearing exactly the same clothes as a dozen of your closest friends. If just two of you were dressed the same, it might be explained away as coincidence, but a dozen suggests that the partygoers had coordinated their attire in advance. In cosmology, the number is not a dozen but tens of thousands—the number of independent yet statistically identical patches of sky in the microwave background.

One possibility is that all those regions of space were endowed at birth with identical properties—in other words, that the homogeneity is mere coincidence. Physicists, however, have thought about two more natural ways out of the impasse: the early universe was much smaller or much older than in standard cosmology. Either (or both, acting together) would have made intercommunication possible.

The most popular choice follows the first alternative. It postulates that the universe went through a period of accelerating expansion, known as inflation, early in its history. Before this phase, galaxies or their precursors were so closely packed that they could easily coordinate their properties. During inflation, they fell out of contact because light was unable to keep pace with the frenetic expansion. After inflation ended, the expansion began to decelerate, so galaxies gradually came back into one another's view.

Physicists ascribe the inflationary spurt to the potential energy stored in a new quantum field, the inflaton, about 10^{-35} second after the big bang. Potential energy, as opposed to rest mass or kinetic energy, leads to gravitational repulsion. Rather than slowing down the expansion, as the gravitation of ordinary matter would, the inflaton accelerated it. Proposed in 1981, inflation has explained a wide variety of observations with precision [see "The Inflationary Universe," by Alan H. Guth and Paul J. Steinhardt; SCIENTIFIC AMERICAN, May 1984; and "Four Keys to Cosmology," Special report; SCIENTIFIC AMERICAN, February]. A number of possible theoretical problems remain, though, beginning with the questions of what exactly the inflaton was and what gave it such a huge initial potential energy.

A second, less widely known way to solve the puzzle follows the second alternative by getting rid of the singularity. If time did not begin at the bang, if a long era preceded the onset of the present cosmic expansion, matter could have had plenty of time to arrange itself smoothly. Therefore, researchers have reexamined the reasoning that led them to infer a singularity.

One of the assumptions—that relativity theory is always valid—is questionable. Close to the putative singularity, quantum effects must have been important, even dominant. Standard relativity takes no account of such effects, so accepting the inevitability of the singularity amounts to trusting the theory beyond reason. To know what really happened, physicists need to subsume relativity in a quantum theory of gravity. The task has occupied theorists from Einstein onward, but progress was almost zero until the mid-1980s.

Evolution of a Revolution

Today two approaches stand out. One, going by the name of loop quantum gravity, retains Einstein's theory essentially intact but changes the procedure for implementing it in quantum mechanics [see "Atoms of Space and Time," by Lee Smolin; SCIENTIFIC AMERICAN, January]. Practitioners of loop quantum gravity have taken great strides and achieved deep insights over the past several years. Still, their approach may not be revolutionary enough to resolve the fundamental problems of quantizing gravity. A similar problem faced particle theorists after Enrico Fermi introduced his effective theory of the weak nuclear force in 1934. All efforts to construct a quantum version of Fermi's theory failed miserably. What was needed was not a new technique but the deep modifications brought by the electroweak theory of Sheldon L. Glashow, Steven Weinberg and Abdus Salam in the late 1960s.

The second approach, which I consider more promising, is string theory—a truly revolutionary modification of Einstein's theory. This article will focus on it, although proponents of loop quantum gravity claim to reach many of the same conclusions.

String theory grew out of a model that I wrote down in 1968 to describe the world of nuclear particles (such as protons and neutrons) and their interactions. Despite much initial excitement, the model failed. It was abandoned several years later in favor of quantum chromodynamics, which describes nuclear particles in terms of more elementary constituents, quarks. Quarks are confined inside a proton or a neutron, as if they were tied together by elastic strings. In retrospect, the original string theory had captured those stringy aspects of the nuclear world. Only later was it revived as a candidate for combining general relativity and quantum theory.

The basic idea is that elementary particles are not pointlike but rather infinitely thin one-dimensional objects, the strings. The large zoo of elementary particles, each with its own characteristic properties, reflects the many possible vibration patterns of a string. How can such a simple-minded theory describe the complicated world of particles and their interactions? The answer can be found in what we may call quantum string magic. Once the rules of quantum mechanics are applied to a vibrating string—just like a miniature violin string, except that the vibrations propagate along it at the speed of light—new properties appear. All have profound implications for particle physics and cosmology.

First, quantum strings have a finite size. Were it not for quantum effects, a violin string could be cut in half, cut in half again and so on all the way down, finally becoming a massless pointlike particle. But the Heisenberg uncertainty principle eventually intrudes and prevents the lightest strings from being sliced smaller than about 10^{-34} meter. This irreducible quantum of length, denoted l_s, is a new constant of nature introduced by string theory side by side with the speed of light, c, and Planck's constant, h. It plays a crucial role in almost every aspect of string theory, putting a finite limit on quantities that otherwise could become either zero or infinite.

Second, quantum strings may have angular momentum even if they lack mass. In classical physics, angular momentum is a property of an object that rotates with respect to an axis. The formula for angular momentum multiplies together velocity, mass and distance from the axis; hence, a massless object can have no angular momentum. But quantum fluctuations change the situation. A tiny string can acquire up to two units of h of angular momentum without gaining any mass. This feature is very welcome because it precisely matches the properties of the carriers of all known fundamental forces, such as the photon (for electromagnetism) and the graviton (for gravity). Historically, angular momentum is what clued in physicists to the quantum-gravitational implications of string theory.

Third, quantum strings demand the existence of extra dimensions of space, in addition to the usual three. Whereas a classical violin string will vibrate no matter what the properties of space and time are, a quantum string is more finicky. The equations describing the vibration become inconsistent unless spacetime either is highly curved (in contradiction with observations) or contains six extra spatial dimensions.

Fourth, physical constants—such as Newton's and Coulomb's constants, which appear in the equations of physics and determine the properties of nature—no longer have arbitrary, fixed values. They occur in string theory as fields, rather like the electromagnetic field, that can adjust their values dynamically. These fields may have taken different values in different cosmological epochs or in remote regions of space, and even today the physical "constants" may vary by a small amount. Observing any variation would provide an enormous boost to string theory.

One such field, called the dilaton, is the master key to string theory; it determines the overall strength of all interactions. The dilaton fascinates string theorists because its value can be reinterpreted as the size of an extra dimension of space, giving a grand total of 11 spacetime dimensions.

Tying Down the Loose Ends

Finally, quantum strings have introduced physicists to some striking new symmetries of nature known as dualities, which alter our intuition for what happens when objects get extremely small. I have already alluded to a form of duality: normally, a short string is lighter than a long one, but if we attempt to squeeze down its size below the fundamental length l_s, the string gets heavier again.

Another form of the symmetry, T-duality, holds that small and large extra dimensions are equivalent. This symmetry arises because strings can move in more complicated ways than pointlike particles can. Consider a closed string (a loop) located on a cylindrically shaped space, whose circular cross section represents one finite extra dimension. Besides vibrating, the string can either turn as a whole around the cylinder or wind around it, one or several times, like a rubber band wrapped around a rolled-up poster.

The energetic cost of these two states of the string depends on the size of the cylinder. The energy of winding is directly proportional to the cylinder radius: larger cylinders require the string to stretch more as it wraps around, so the windings contain more energy than they would on a smaller cylinder. The energy associated with moving around the circle, on the other hand, is inversely proportional to the radius: larger cylinders allow for longer wavelengths (smaller frequencies), which represent less energy than shorter wavelengths do. If a large cylinder is substituted for a small one, the two states of motion can swap roles. Energies that had been produced by circular motion are instead produced by winding, and vice versa. An outside observer notices only the energy levels, not the origin of those levels. To that observer, the large and small radii are physically equivalent.

Although T-duality is usually described in terms of cylindrical spaces, in which one dimension (the circumference) is finite, a variant of it applies to our ordinary three dimensions, which appear to stretch on indefinitely. One must be careful when talking about the expansion of an infinite space. Its overall size cannot change; it remains infinite. But it can still expand in the sense that bodies embedded within it, such as galaxies, move apart from one another. The crucial variable is not the size of the space as a whole but its scale factor—the factor by which the distance between galaxies changes, manifesting itself as the galactic redshift that astronomers observe. According to T-duality, universes with small scale factors are equivalent to ones with large scale factors. No such symmetry is present in Einstein's equations; it emerges from the unification that string theory embodies, with the dilaton playing a central role.

For years, string theorists thought that T-duality applied only to closed strings, as opposed to open strings, which have loose ends and thus cannot wind. In 1995 Joseph Polchinski of the University of California at Santa Barbara realized that T-duality did apply to open strings, provided that the switch between large and small radii was accompanied by a change in the conditions at the end points of the string. Until then, physicists had postulated boundary conditions in which no force acted on the ends of the strings, leaving them free to flap around. Under

T-duality, these conditions become so-called Dirichlet boundary conditions, whereby the ends stay put.

Any given string can mix both types of boundary conditions. For instance, electrons may be strings whose ends can move around freely in three of the 10 spatial dimensions but are stuck within the other seven. Those three dimensions form a subspace known as a Dirichlet membrane, or D-brane. In 1996 Petr Horava of the University of California at Berkeley and Edward Witten of the Institute for Advanced Study in Princeton, N.J., proposed that our universe resides on such a brane. The partial mobility of electrons and other particles explains why we are unable to perceive the full 10-dimensional glory of space.

Taming the Infinite

All the magic properties of quantum strings point in one direction: strings abhor infinity. They cannot collapse to an infinitesimal point, so they avoid the paradoxes that collapse entails. Their nonzero size and novel symmetries set upper bounds to physical quantities that increase without limit in conventional theories, and they set lower bounds to quantities that decrease. String theorists expect that when one plays the history of the universe backward in time, the curvature of spacetime starts to increase. But instead of going all the way to infinity (at the traditional big bang singularity), it eventually hits a maximum and shrinks once more. Before string theory, physicists were hard-pressed to imagine any mechanism that could so cleanly eliminate the singularity.

Conditions near the zero time of the big bang were so extreme that no one yet knows how to solve the equations. Nevertheless, string theorists have hazarded guesses about the pre-bang universe. Two popular models are floating around.

The first, known as the pre-big bang scenario, which my colleagues and I began to develop in 1991, combines T-duality with the better-known symmetry of time reversal, whereby the equations of physics work equally well when applied backward and forward in time. The combination gives rise to new possible cosmologies in which the universe, say, five seconds before the big bang expanded at the same pace as it did five seconds after the bang. But the rate of change of the expansion was opposite at the two instants: if it was decelerating after the bang, it was accelerating before. In short, the big bang may not have been the origin of the universe but simply a violent transition from acceleration to deceleration.

The beauty of this picture is that it automatically incorporates the great insight of standard inflationary theory—namely, that the universe had to undergo a period of acceleration to become so homogeneous and isotropic. In the standard theory, acceleration occurs after the big bang because of an ad hoc inflaton field. In the pre-big bang scenario, it occurs before the bang as a natural outcome of the novel symmetries of string theory.

According to the scenario, the pre-bang universe was almost a perfect mirror image of the post-bang one. If the universe is eternal into the future, its contents thinning to a meager gruel, it is also eternal into the past. Infinitely long ago it was nearly empty, filled only with a tenuous, widely dispersed, chaotic gas of radiation and matter. The forces of nature, controlled by the dilaton field, were so feeble that particles in this gas barely interacted.

As time went on, the forces gained in strength and pulled matter together. Randomly, some regions accumulated matter at the expense of their surroundings. Eventually the density in these regions became so high that black holes started to form. Matter inside those regions was then cut off from the outside, breaking up the universe into disconnected pieces.

Inside a black hole, space and time swap roles. The center of the black hole is not a point in space but an instant in time. As the infalling matter approached the center, it reached higher and higher densities. But when the density, temperature and curvature reached the maximum values allowed by string theory, these quantities bounced and started decreasing. The moment of that reversal is what we call a big bang. The interior of one of those black holes became our universe.

Not surprisingly, such an unconventional scenario has provoked controversy. Andrei Linde of Stanford University has argued that for this scenario to match observations, the black hole that gave rise to our universe would have to have formed with an unusually large size—much larger than the length scale of string theory. An answer to this objection is that the equations predict black holes of all possible sizes. Our universe just happened to form inside a sufficiently large one.

A more serious objection, raised by Thibault Damour of the Institut des Hautes Études Scientifiques in Bures-sur-Yvette, France, and Marc Henneaux of the Free University of Brussels, is that matter and spacetime would have behaved chaotically near the moment of the bang, in possible contradiction with the observed regularity of the early universe. I have recently proposed that a chaotic state would produce a dense gas of miniature "string holes"—strings that were so small and massive that they were on the verge of becoming black holes. The behavior of these holes could solve the problem identified by Damour and Henneaux. A similar proposal has been put forward by Thomas Banks of Rutgers University and Willy Fischler of the University of Texas at Austin. Other critiques also exist, and whether they have uncovered a fatal flaw in the scenario remains to be determined.

Bashing Branes

The other leading model for the universe before the bang is the ekpyrotic ("conflagration") scenario. Developed three years ago by a team of cosmologists and string theorists—Justin Khoury of Columbia

University, Paul J. Steinhardt of Princeton University, Burt A. Ovrut of the University of Pennsylvania, Nathan Seiberg of the Institute for Advanced Study and Neil Turok of the University of Cambridge—the ekpyrotic scenario relies on the idea that our universe is one of many D-branes floating within a higher-dimensional space. The branes exert attractive forces on one another and occasionally collide. The big bang could be the impact of another brane into ours.

In a variant of this scenario, the collisions occur cyclically. Two branes might hit, bounce off each other, move apart, pull each other together, hit again, and so on. In between collisions, the branes behave like Silly Putty, expanding as they recede and contracting somewhat as they come back together. During the turnaround, the expansion rate accelerates; indeed, the present accelerating expansion of the universe may augur another collision.

The pre-big bang and ekpyrotic scenarios share some common features. Both begin with a large, cold, nearly empty universe, and both share the difficult (and unresolved) problem of making the transition between the pre- and the post-bang phase. Mathematically, the main difference between the scenarios is the behavior of the dilaton field. In the pre-big bang, the dilaton begins with a low value—so that the forces of nature are weak—and steadily gains strength. The opposite is true for the ekpyrotic scenario, in which the collision occurs when forces are at their weakest.

The developers of the ekpyrotic theory initially hoped that the weakness of the forces would allow the bounce to be analyzed more easily, but they were still confronted with a difficult high-curvature situation, so the jury is out on whether the scenario truly avoids a singularity.

Also, the ekpyrotic scenario must entail very special conditions to solve the usual cosmological puzzles. For instance, the about-to-collide branes must have been almost exactly parallel to one another, or else the collision could not have given rise to a sufficiently homogeneous bang. The cyclic version may be able to take care of this problem, because successive collisions would allow the branes to straighten themselves.

Leaving aside the difficult task of fully justifying these two scenarios mathematically, physicists must ask whether they have any observable physical consequences. At first sight, both scenarios might seem like an exercise not in physics but in metaphysics—interesting ideas that observers could never prove right or wrong. That attitude is too pessimistic. Like the details of the inflationary phase, those of a possible pre-bangian epoch could have observable consequences, especially for the small variations observed in the cosmic microwave background temperature.

First, observations show that the temperature fluctuations were shaped by acoustic waves for several hundred thousand years. The regularity of the fluctuations indicates that the waves were synchronized.

Cosmologists have discarded many cosmological models over the years because they failed to account for this synchrony. The inflationary, pre-big bang and ekpyrotic scenarios all pass this first test. In these three models, the waves were triggered by quantum processes amplified during the period of accelerating cosmic expansion. The phases of the waves were aligned.

Second, each model predicts a different distribution of the temperature fluctuations with respect to angular size. Observers have found that fluctuations of all sizes have approximately the same amplitude. (Discernible deviations occur only on very small scales, for which the primordial fluctuations have been altered by subsequent processes.) Inflationary models neatly reproduce this distribution. During inflation, the curvature of space changed relatively slowly, so fluctuations of different sizes were generated under much the same conditions. In both the stringy models, the curvature evolved quickly, increasing the amplitude of small-scale fluctuations, but other processes boosted the large-scale ones, leaving all fluctuations with the same strength. For the ekpyrotic scenario, those other processes involved the extra dimension of space, the one that separated the colliding branes. For the pre-big bang scenario, they involved a quantum field, the axion, related to the dilaton. In short, all three models match the data.

Third, temperature variations can arise from two distinct processes in the early universe: fluctuations in the density of matter and rippling caused by gravitational waves. Inflation involves both processes, whereas the pre-big bang and ekpyrotic scenarios predominantly involve density variations. Gravitational waves of certain sizes would leave a distinctive signature in the polarization of the microwave background [see "Echoes from the Big Bang," by Robert R. Caldwell and Marc Kamionkowski; SCIENTIFIC AMERICAN, January 2001]. Future observatories, such as European Space Agency's Planck satellite, should be able to see that signature, if it exists—providing a nearly definitive test.

A fourth test pertains to the statistics of the fluctuations. In inflation the fluctuations follow a bell-shaped curve, known to physicists as a Gaussian. The same may be true in the ekpyrotic case, whereas the pre-big bang scenario allows for sizable deviation from Gaussianity.

Analysis of the microwave background is not the only way to verify these theories. The pre-big bang scenario should also produce a random background of gravitational waves in a range of frequencies that, though irrelevant for the microwave background, should be detectable by future gravitational-wave observatories. Moreover, because the pre-big bang and ekpyrotic scenarios involve changes in the dilaton field, which is coupled to the electromagnetic field, they would both lead to large-scale magnetic field fluctuations. Vestiges of these fluctuations might show up in galactic and intergalactic magnetic fields.

So, when did time begin? Science does not have a conclusive answer yet, but at least two potentially testable theories plausibly hold that the universe—and therefore time—existed well before the big bang. If either scenario is right, the cosmos has always been in existence and, even if it recollapses one day, will never end.

2 Dark Energy and the Accelerating Universe

In the next article, Saul Perlmutter describes how two teams of scientists made the startling discovery, announced in 1998, that the universe's expansion has increased over time, contrary to the expectation that the gravity of the universe's overall mass would slow it down.

Perlmutter's group used type Ia supernovas as "standard candles" to track the cosmic expansion's change over time. These supernovas exhibit an "amazing consistency" in both spectral emissions and in the waxing and waning of brightness in the weeks following a supernova explosion.

This consistency allows using the observed brightness to estimate a supernova's distance. Combining this with redshift, which indicates how fast the supernova is receding from us due to expansion, gave Perlmutter and his colleagues the tool they needed.

The results from analyzing several dozen measurements stunned the researchers, who found that the oldest supernovas were farther *away than expected. It could only mean that something counters gravitational slowdown. Restoring an antiexpansion term to Einstein's general theory of relativity, which Einstein yanked after Edwin Hubble discovered expansion, allowed cosmologists to determine that about 70 percent of the universe's energy density is in the form of this previously unknown "dark energy," and the rest in matter. —RA*

"Supernovae, Dark Energy, and the Accelerating Universe"
by Saul Perlmutter
***Physics Today*, April 2003**

Using very distant supernovae as standard candles, one can trace the history of cosmic expansion and try to find out what's currently speeding it up.

For millennia, cosmology has been a theorist's domain, where elegant theory was only occasionally endangered by inconvenient facts. Early in the 20th century, Albert Einstein gave us new conceptual tools to rigorously address the questions of the origins, evolution, and fate of the universe. In recent years, technology has developed to the point where these concepts from general relativity can be substantiated and elaborated by measurements. For example, measurement of the remnant glow from the hot, dense beginnings of the expanding universe—the cosmic microwave background—is yielding increasingly detailed data about the first half-million years and the overall geometry of the cosmos.

The standard model of particle physics has also begun to play a prominent role in cosmology. The widely accepted idea of exponential inflation in the immediate aftermath of the Big Bang was built on the predicted effect of certain putative particle fields and potentials on the cosmic expansion. Measuring the history of cosmic expansion is no easy task, but in recent years, a specific variety of supernovae, type Ia, has given us a first glimpse at that history—and surprised us with an unexpected plot twist.

Searching for a Standard Candle

In principle, the expansion history of the cosmos can be determined quite easily, using as a "standard candle" any distinguishable class of astronomical objects of known intrinsic brightness that can be identified over a wide distance range. As the light from such beacons travels to Earth through an expanding universe, the cosmic expansion stretches not only the distances between galaxy clusters, but also the very wavelengths of the photons en route. By the time the light reaches us, the spectral wavelength λ has thus been redshifted by precisely the same incremental factor $z \equiv \Delta\lambda/\lambda$ by which the cosmos has been stretched in the time interval since the light left its source. That time interval is the speed of light times the object's distance from Earth, which can be determined by comparing its apparent brightness to a nearby standard of the same class of astrophysical objects.

The recorded redshift and brightness of each such object thus provide a measurement of the total integrated expansion of the universe since the time the light was emitted. A collection of such measurements, over a sufficient range of distances, would yield an entire historical record of the universe's expansion.

Conceptually, this scheme is a remarkably straightforward means to a profound prize: an empirical account of the growth of our universe. A spectroscopically distinguishable class of objects with determinable intrinsic brightness would do the trick. In Edwin Hubble's discovery of the cosmic expansion in the 1920s, he used entire galaxies as standard candles. But galaxies, coming in many shapes and sizes, are difficult to match against a standard brightness. They can grow fainter with time, or brighter—by merging with other galaxies. In the 1970s, it was suggested that the brightest member of a galaxy cluster might serve as a reliable standard candle. But in the end, all proposed distant galactic candidates were too susceptible to evolutionary change.

As early as 1938, Walter Baade, working closely with Fritz Zwicky, pointed out that supernovae were extremely promising candidates for measuring the cosmic expansion. Their peak brightness seemed to be quite uniform, and they were bright enough to be seen at extremely large distances.[1] In fact, a supernova can, for a few weeks, be as bright as an entire galaxy. Over the years, however, as more and more supernovae were measured, it became clear that they were a rather heterogeneous group with a wide range of intrinsic peak brightnesses.

In the early 1980s, a new subclassification of supernovae emerged. Supernovae with no hydrogen features in their spectra had previously all been classified simply as type I. Now this class was subdivided into types Ia and Ib, depending on the presence or absence of a silicon absorption feature at 6150 Å in the supernova's spectrum.[2] With that minor improvement in typology, an amazing consistency among the type Ia supernovae became evident. Their spectra matched feature-by-feature, as did their "light curves"—the plots of waxing and waning brightness in the weeks following a supernova explosion.[3, 4]

The uniformity of the type Ia supernovae became even more striking when their spectra were studied in detail as they brightened and then faded. First, the outermost parts of the exploding star emit a spectrum that's the same for all typical type Ia supernovae, indicating the same elemental densities, excitation states, velocities, and so forth. Then, as the exploding ball of gas expands, the outermost layers thin out and become transparent, letting us see the spectral signatures of conditions further inside. Eventually, if we watch the entire time series of spectra, we get to see indicators that probe almost the entire explosive event. It is impressive that the type Ia supernovae exhibit so much uniformity down to this level of detail. Such a "supernova CAT-scan" can be difficult to interpret. But it's clear that essentially the same physical processes are occurring in all of these explosions.

The detailed uniformity of the type Ia supernovae implies that they must have some common triggering mechanism. Equally important, this uniformity provides standard spectral and light-curve templates that offer the possibility of singling out those supernovae that deviate slightly from the norm. The complex natural histories of galaxies had made them difficult to standardize. With type Ia supernovae, however, we saw the chance to avoid such problems. We could examine the rich stream of observational data from each individual explosion and match spectral and light-curve fingerprints to recognize those that had the same peak brightness.

Within a few years of their classification, type Ia supernovae began to bear out that expectation. First, David Branch and coworkers at the University of Oklahoma showed that the few type Ia outliers—those with peak brightness significantly different from the norm—could generally be identified and screened out.[4] Either their spectra or their "colors" (the ratios of intensity seen through two broadband filters) deviated from the templates. The anomalously fainter supernovae were typically redder or found in highly inclined spiral galaxies (or both). Many of these were presumably dimmed by dust, which absorbs more blue light than red.

Soon after Branch's work, Mark Phillips at the Cerro Tololo Interamerican Observatory in Chile showed that the type Ia brightness outliers also deviated from the template light curve—and in a very predictable way.[5] The supernovae that faded faster than the norm were fainter at their peak, and the slower ones were brighter. In fact, one could use the light curve's time scale to predict peak brightness and thus slightly recalibrate each supernova. But the great majority of type Ia supernovae, as Branch's group showed, passed the screening tests and were, in fact, excellent standard candles that needed no such recalibration.[6]

Cosmological Distances

When the veteran Swiss researcher Gustav Tammann and his student Bruno Leibundgut first reported the amazing uniformity of type Ia supernovae, there was immediate interest in trying to use them to determine the Hubble constant, H_0, which measures the present expansion rate of the cosmos. That could be done by finding and measuring a few type Ia supernovae just beyond the nearest clusters of galaxies, that is, explosions that occurred some 100 million years ago. An even more challenging goal lay in the tantalizing prospect that we could find such standard-candle supernovae more than ten times farther away and thus sample the expansion of the universe several billion years ago. Measurements using such remote supernovae might actually show the expected slowing of the expansion rate by gravity. Because that deceleration rate would depend on the cosmic mean mass density ρ_m, we would, in effect, be weighing the universe.

If mass density is, as was generally supposed a decade ago, the primary energy constituent of the universe, then the measurement of the changing expansion rate would also determine the curvature of space and tell us about whether the cosmos is finite or infinite. Furthermore, the fate of the universe might be said to hang in the balance: If, for example, we measured a cosmic deceleration big enough to imply a ρ_m exceeding the "critical density" ρ_c (roughly 10^{-29} gm/cm^3), that would indicate that the universe will someday stop expanding and collapse toward an apocalyptic "Big Crunch."

All this sounded enticing: fundamental measurements made with a new distance standard bright enough to be seen at cosmological distances. The problem was that type Ia supernovae are a pain in the neck, to be avoided if anything else would do. At the time, a brief catalog of reasons *not* to pursue cosmological measurement with type Ia supernovae might have begun like this:

- They are rare. A typical galaxy hosts only a couple of type Ia explosions per millennium.

- They are random, giving no advance warning of where to look. But the scarce observing time at the world's largest telescopes, the only tools powerful enough to measure these most distant supernovae adequately, is allocated on the basis of research proposals written more than six months in advance. Even the few successful proposals are granted only a few nights per semester. The possible occurrence of a chance supernova doesn't make for a compelling proposal.

- They are fleeting. After exploding, they must be discovered promptly and measured multiple times within a few weeks, or they will already have passed the peak brightness that is essential for calibration. It's too late to submit the observing proposal after you've discovered the supernova.

This was a classic catch-22. You couldn't preschedule telescope time to identify a supernova's type or follow it up if you couldn't guarantee one. But you couldn't prove a technique for guaranteeing type Ia supernova discoveries without prescheduling telescope time to identify them spectroscopically.

The list of problems didn't stop there. The increasing redshifting of supernova spectra with distance means that the brightness of a very distant supernova measured through a given filter is hard to compare with the brightness of a much closer supernova measured through the same filter. (Astronomers call this the K-correction problem.) Dust in a supernova's host galaxy can dim the explosion's light. And there were doubts that the spectra of faint distant supernovae could be reliably identified as type Ia.

In fact, the results from the first search for very distant type Ia supernovae were not encouraging. In the late 1980s, a Danish team led by Hans Nørgaard-Nielsen found only one type Ia supernova in two years of intensive observing, and that one was already several weeks past its peak.

A Systematic Solution

Daunting as these problems appeared, it seemed crazy to let the logistics stand in the way, when the tools were at hand for measuring such fundamental properties of the universe: its mass density, its large-scale curvature, and its fate. After all, we didn't have to build anything nearly as formidable as the gargantuan accelerators and detectors needed for particle physics. In a project that

Carl Pennypacker and I began in Richard Muller's group at the University of California, Berkeley, just before the Danish team's 1988 supernova discovery, we started by building a wide-field imager for the Anglo-Australian Observatory's 4-meter telescope. The imager would let us study thousands of distant galaxies in a night, upping the odds of a supernova discovery. Contemporary computing and networking advances just barely made possible the next-day analysis that would let us catch supernovae as they first brightened.

Finding our first supernova in 1992, we also found a solution to the K-correction problem by measuring the supernova in a correspondingly redshifted filter. By playing this trick with two redshifted filter bands, one could also expect to recognize dust absorption by its wavelength dependence. But we still hadn't solved the catch-22 telescope scheduling problem. We couldn't preschedule follow-up observations of our first supernova, so couldn't obtain its identifying spectrum.

In retrospect, the solution we found seems obvious —though much effort was needed to implement it and prove it practical. By specific timing of the requested telescope schedules, we could guarantee that our wide-field imager would harvest a batch of about a dozen freshly exploded supernovae, all discovered on a pre-specified observing date during the dark phase of the moon. (A bright moon is an impediment to the follow-up observation.) We first demonstrated this supernovae-on-demand methodology in 1994. From then on, our proposals for time at major ground-based telescopes could specify discovery dates and roughly how many supernovae would be found and followed up. This approach also made it possible to use the Hubble Space Telescope for follow-up light-curve observations, because we could specify in advance the one-square-degree patch of sky in which our wide-field imager would find its catch of supernovae. Such specificity is a requirement for advance scheduling of the HST. By now, the Berkeley team, had grown to include some dozen collaborators around the world, and was called the Supernova Cosmology Project (SCP).

A Community Effort

Meanwhile, the whole supernova community was making progress with the understanding of relatively nearby supernovae. Mario Hamuy and coworkers at Cerro Tololo took a major step forward by finding and studying many nearby (low-redshift) type Ia supernovae.[7] The resulting beautiful data set of 38 supernova light curves made it possible to check and improve on the results of Branch and Phillips, showing that type Ia peak brightness could be standardized.[6, 7]

The new supernovae-on-demand techniques that permitted systematic study of distant supernovae and the improved understanding of brightness variations among nearby type Ia's spurred the community to redouble its efforts. A second collaboration, called the High-Z Supernova Search and led by Brian Schmidt of Australia's Mount Stromlo Observatory, was formed at the end of 1994. The team includes many veteran supernova experts. The two rival teams raced each other over the next few years—occasionally covering for each other with observations when one of us had bad weather—as we all worked feverishly to find and study the guaranteed on-demand batches of supernovae.

At the beginning of 1997, the SCP team presented the results for our first seven high-redshift supernovae.[8] These first results demonstrated the cosmological analysis techniques from beginning to end. They were suggestive of an expansion slowing down at about the rate expected for the simplest inflationary Big Bang models, but with error bars still too large to permit definite conclusions.

By the end of the year, the error bars began to tighten, as both groups now submitted papers with a few more supernovae, showing evidence for much less than the expected slowing of the cosmic expansion.[9-11] This was beginning to be a problem for the simplest inflationary models with a universe dominated by its mass content.

Finally, at the beginning of 1998, the two groups presented the results shown in figure 3 [in the original article].[12, 13]

What's Wrong with Faint Supernovae?

The faintness—or distance—of the high-redshift supernovae in figure 3 [in the original article] was a dramatic surprise. In the simplest cosmological models, the expansion history of the cosmos is determined entirely by its mass density. The greater the density, the more the expansion is slowed by gravity. Thus, in the past, a high-mass-density universe would have been expanding much faster than it does today. So one shouldn't have to look far back in time to especially distant (faint) supernovae to find a given integrated expansion (redshift).

Conversely, in a low-mass-density universe one would have to look farther back. But there is a limit to how low the mean mass density could be. After all, we are here, and the stars and galaxies are here. All that mass surely puts a lower limit on how far—that is, to what level of faintness—we must look to find a given redshift. The high redshift supernovae in figure 3 [in the original article] are, however, fainter than would be expected *even for an empty cosmos.*

If these data are correct, the obvious implication is that the simplest cosmological model must be too simple. The next simplest model might be one that Einstein entertained for a time. Believing the universe to be static, he tentatively introduced into the equations of general relativity an expansionary term he called the "cosmological constant" (Λ) that would compete against gravitational collapse. After Hubble's discovery of the cosmic expansion, Einstein famously rejected Λ as his "greatest blunder." In later years, Λ came to be identified with the zero-point vacuum energy of all quantum fields.

It turns out that invoking a cosmological constant allows us to fit the supernova data quite well. (Perhaps there was more insight in Einstein's blunder than in the best efforts of ordinary mortals.) In 1995, my SCP colleague Ariel Goobar and I had found that, with a sample of type Ia supernovae spread over a sufficiently wide range of distances, it would be possible to separate out the competing effects of the mean mass density and the vacuum-energy density.[14]

The best fit to the 1998 supernova data implies that, in the present epoch, the vacuum energy density ρ_Λ is larger than the energy density attributable to mass $(\rho_m c^2)$. Therefore, the cosmic expansion is now accelerating. If the universe has no large-scale curvature, as the recent measurements of the cosmic microwave background strongly indicate, we can say quantitatively that about 70 % of the total energy density is vacuum energy and 30 % is mass. In units of the critical density ρ_c, one usually writes this result as

$$\Omega_\Lambda \equiv \rho_\Lambda/\rho_c \approx 0.7 \text{ and } \Omega_m \equiv \rho_m/\rho_c \approx 0.3.$$

Why not a Cosmological Constant?

The story might stop right here with a happy ending— a complete physics model of the cosmic expansion—were it not for a chorus of complaints from the particle theorists. The standard model of particle physics has no natural place for a vacuum energy density of the modest magnitude required by the astrophysical data. The simplest estimates would predict a vacuum energy 10^{120} times greater. (In supersymmetric models, it's "only" 10^{55} times greater.) So enormous a Λ would have engendered an acceleration so rapid that stars and galaxies could never have formed. Therefore it has long been assumed that there must be some underlying symmetry that precisely cancels the vacuum energy. Now, however, the supernova data appear to require that such a cancellation would have to leave a remainder of about one part in 10^{120}. That degree of fine tuning is most unappealing.

The cosmological constant model requires yet another fine tuning. In the cosmic expansion, mass density becomes ever more dilute. Since the end of inflation, it has fallen by very many orders of magnitude. But the vacuum energy density ρ_Λ, a property of empty space itself, stays constant. It seems a remarkable and implausible coincidence that the mass density, just in the present epoch, is within a factor of 2 of the vacuum energy density.

Given these two fine-tuning coincidences, it seems likely that the standard model is missing some fundamental physics. Perhaps we need some new kind of accelerating energy—a "dark energy" that, unlike Λ, is not constant. Borrowing from the example of the putative "inflaton" field that is thought to have triggered inflation, theorists are proposing dynamical scalar-field models and other even more exotic alternatives to a cosmological constant, with the goal of solving the coincidence problems.

The experimental physicist's life, however, is dominated by more prosaic questions: "Where could my measurement be wrong, and how can I tell?" Crucial questions of replicability were answered by the striking agreement between our results and those of the competing team, but there remain the all-important questions of systematic uncertainties. Most of the two groups' efforts have been devoted to hunting down these systematics.[15, 16] Could the faintness of the supernovae be due to intervening dust? The color measurements that would show color-dependent dimming for most types of dust indicate that dust is not a major factor.[12, 13] Might the type Ia supernovae have been intrinsically fainter in the

86

distant past? Spectral comparisons have, thus far, revealed no distinction between the exploding atmospheres of nearby and more distant supernovae.[9, 12]

Another test of systematics is to look for even more distant supernovae, from the time when the universe was so much more dense that ρ_m dominated over the dark energy and was thus still slowing the cosmic expansion. Supernovae from that *decelerating* epoch should not get as faint with increasing distance as they would if dust or intrinsic evolutionary changes caused the dimming. The first few supernovae studied at redshifts beyond $z = 1$ have already begun to constrain these systematic uncertainties.[17] (See PHYSICS TODAY, June 2001, page 17.)

By confirming the flat geometry of the cosmos, the recent measurements of the cosmic microwave background have also contributed to confidence in the accelerating-universe results. Without the extra degree of freedom provided by possible spatial curvature, one would have to invoke improbably large systematic error to negate the supernova results. And if we include the low ρ_m estimates based on inventory studies of galaxy clusters, the $\Omega_m - \Omega_\Lambda$ parameter plane shows a reassuring overlap for the three independent kinds of cosmological observations.

Pursuing the Elusive Dark Energy

The dark energy evinced by the accelerating cosmic expansion grants us almost no clues to its identity. Its tiny density and its feeble interactions presumably preclude identification in the laboratory. By construction, of course, it does affect the expansion rate of the universe, and different dark-energy models imply different expansion rates in different epochs. So we must hunt for the fingerprints of dark energy in the fine details of the history of cosmic expansion.

The wide-ranging theories of dark energy are often characterized by their equation-of-state parameter $w \equiv p/\rho$, the ratio of the dark energy's pressure to its energy density. The deceleration (or acceleration) of an expanding universe, given by the general relativistic equation

$$\ddot{R}/R = -4/3\pi G\rho\ (1\ +\ 3w),$$

depends on this ratio. Here R, the linear scale factor of the expanding universe, can be thought of as the mean distance between galaxy clusters not bound to each other. Thus the expansion accelerates whenever w is more negative than $-1/3$, after one includes all matter, radiation, and dark-energy components of the cosmic energy budget.

Each of the components has its own w: negligible for nonrelativistic matter, $+1/3$ for radiation and relativistic matter, and -1 for Λ. That is, Λ exerts a peculiar negative pressure! General relativity also tells us that each component's energy density falls like $R^{-3(1\ +\ w)}$ as the cosmos expands. Therefore, radiation's contribution falls away first, so that nonrelativistic matter and dark energy now predominate. Given that the dark-energy density is now about twice the mass density, the only constraint on dark-energy models is that w must, at present, be more negative than $-1/2$ to make the cosmic expansion accelerate. However, most dark-energy alternatives to a cosmological constant have a w that changes over time. If we can learn more about the history of cosmic expansion, we can hope to discriminate among theories of dark energy by better determining w and its time dependence.

Unfortunately, the differences between the expansion histories predicted by the current crop of dark-energy models are extremely small. Distinguishing among them will require measurements an order of magnitude more accurate than those shown in figure 3 [in the original article], and extending twice as far back in time.

There is no shortage of type Ia supernovae; one explodes somewhere in the sky every few seconds. In principle, then, the job is simply to study a hundred times as many supernovae as we have so far. That's a difficult but not prohibitive task, if we install dedicated wider-field imagers and improved spectrographs on dedicated large telescopes. However, it's not just a matter of improving the quantity of measurements. The quality must also take a dramatic step forward, because the current measurement accuracy is not limited simply by statistical errors. Even with the number of supernovae we already have in hand, our statistical uncertainties are already close to the systematic uncertainties.

A New Challenge

The next generation of supernova projects has already begun. Telescope scheduling committees have dramatically increased the time allotted them on the largest telescopes. With biweekly monitoring of patches of sky for several years on end at two 4-meter telescopes, it will be possible to collect almost complete light curves for hundreds of 5-billion-year-old type Ia supernovae. Smaller telescopes will study the time-varying spectra of much closer supernovae, and imagers on the HST and the 8-m Subaru Telescope in Hawaii are now revealing handfuls of 10-billion-year-old supernovae. A number of large new telescopes are devoting extensive observing programs to follow-up measurements of this plethora of supernovae. At the most extreme distances, only the Hubble telescope can just barely follow the fading supernovae, red-shifted into the infrared. With this array of effort, we may know, before too long, whether the time-averaged behavior of the dark energy is consistent with a cosmological constant.

The still harder goal of the third generation of supernova work, which also has already begun, is to look for time variations in the dark energy. For this higher-precision work, the systematic uncertainties must be reduced dramatically. The physical details of each individual supernova explosion must be pinned down with extensive and exacting spectral and photometric monitoring. Intervening dust must be measured with wavelength coverage extending into the near-infrared. Host galaxies must be classified to control for environmental effects on the type Ia standard candle. And we will have to study enough supernovae in each redshift range to take account of possible gravitational lensing by foreground galaxies that can brighten or dim a supernova.

These very exacting requirements have pushed us to work above the atmosphere and design a new orbiting optical and near-infrared telescope called SNAP (the SuperNova/Acceleration Probe). With a 2-meter mirror, a half-billion-pixel imager, and a high-throughput spectrograph, this space mission can accomplish the unprecedented suite of measurements required for measuring thousands of supernovae with adequately constrained systematic uncertainties.[18]

We live in an unusual time, perhaps the first golden age of empirical cosmology. With advancing technology, we have begun to make philosophically significant measurements. These measurements have already brought surprises. Not only is the universe accelerating, but it apparently consists primarily of mysterious substances. We've already had to revise our simplest cosmological models. Dark energy has now been added to the already perplexing question of dark matter. One is tempted to speculate that these ingredients are add-ons, like the Ptolemaic epicycles, to preserve an incomplete theory. With the next decade's new experiments, exploiting not only distant supernovae, but also the cosmic microwave background, gravitational lensing of galaxies, and other cosmological observations, we have the prospect of taking the next step toward that "Aha!" moment when a new theory makes sense of the current puzzles.

References

1. W. Baade, *Astrophys. J.* 88, 285 (1938); C. Kowal, *Astron. J.* 73, 1021 (1968).

2. J. C. Wheeler, R. Levreault, *Astrophys. J. Lett.* 294, 17 (1985); A. Uomoto, R. Kirshner, *Astron. Astrophys.* 149, L7 (1985); N. Panagia, in *Supernovae as Distance Indicators*, N. Bartel, ed., Springer-Verlag, Berlin (1985); R. Harkness, J. C. Wheeler, in *Supernovae*, A. Petschek, ed., Springer-Verlag, New York (1990).

3. B. Leibungdgut, PhD Thesis, University of Basel (1988); G. Tammann, B. Leibundgut, *Astron. Astrophys.* 236, 9 (1990).

4. D. Branch, G. Tammann, *Annu. Rev. Astron. Astrophys.* 30, 359 (1992); D. Miller, D. Branch, *Astron. J.* 100, 530 (1990); D. Branch, D. Miller, *Astrophys. J. Lett.* 495, L5 (1993).

5. M. Phillips, *Astrophys. J. Lett.* 413, 105, (1993); A. Riess, W. Press, R. Kirshner, *Astrophys. J.* 47388, (1996). See also ref. 8.

6. D. Branch, A. Fisher, P. Nugent, *Astron. J.* 106, 2383 (1993); T. Vaughan, D. Branch, D. Miller, S. Perlmutter, *Astrophys. J.* 439, 5S8 (1995).

7. M. Hamuy et al., *Astron. J.* 106, 2392 (1993); and 109, 1 (1995).

8. S. Perlmutter et al. (Supernova Cosmology Project), *Astrophys. J.* 483, 565 (1997).

9. S. Perlmutter et al. (Supernova Cosmology Project), *Nature* 391, 51 (1998).

10. P. Garnavich et al. (High-Z Supernova Search), *Astrophys. J. Lett.* 493, 53 (1998).

11. B. Schmidt et al. (High-Z Supernova Search), *Astrophys. J.* 507, 46 (1998).

12. A. Riess, A. Filippenko, P. Challis, A. Clocchiattia, A. Diercks, P. Garnavich, R. Gilliland, C. Hogan, S. Jha, R. Kirshner, B. Leibundgut, M. Phillips, D. Reiss, B. Schmidt, R. Schommer, R. Smith, J. Spyromilio, C. Stubbs, N. Suntzeff, J. Tonry, *Astron. J.* 116, 1009 (1998).

13. S. Perlmutter, G. Aldering, G. Goldhaber, R. Knop, P. Nugent, P. Castro, S. Deustua, S. Fabbro, A. Goobar, D. Groom, I. Hook, A. Kim, M. Kim, J. Lee, N. Nunes, R. Pain, C. Pennypacker, R. Quimby, C. Lidman, R. Ellis, M. Irwin, R. McMahon, P. Ruiz-Lapuente, N. Walton, B. Schaefer, B. Boyle, A. Filippenko, T. Matheson, A. Fruchter, N. Panagia, H. Newberg, W. Couch, *Astrophys. J.* 517, 565, (1999).

14. A. Goobar, S. Perlmutter, *Astrophys. J.* 450, 14 (1995).

15. S. Perlmutter, B. Schmidt, in *Supernovae and Gamma Ray Bursters*, K. Weiler, ed., Springer-Verlag, New York (2003), and references therein. Available at http:/xxx.lanl.gov/astro-ph/0303428.

16. See the Web sites of the Supernova Cosmology Project, http://supernova. LBL.gov, and the High-Z Supernova Search, http://cfa-www.harvard.edu/cfa/ oir/Research/supernova/ HighZ.html.

17. See, for example, A. Riess et al., *Astrophys. J.*, 560, 49 (2001).

18. For more information on SNAP, see http://snap.lbl.gov.

Reprinted with permission from "Supernovae, Dark Energy, and the Accelerating Universe," by Perlmutter, Saul, *Physics Today*, April 2003, pp. 53–60. © 2003, American Institute of Physics.

Although the 1998 discovery of the accelerating universe suggested the presence of "dark energy" that opposes gravity, it took another five years of using a highly detailed picture of the cosmic microwave background (CMB) to satisfy lingering doubts about its existence.

The CMB is the "most ancient light in the universe," author Charles Seife writes in the next article. It's the radiation that "streamed from the newborn universe when it was still a glowing ball of plasma."

In 2003, the Wilkinson Microwave Anisotropy Probe (WMAP) satellite produced a high-resolution image of the "infant cosmos." The image bears the faint imprint of acoustic reverberations in the plasma, with hot spots showing denser, compressed regions. These subtle variations allowed scientists to nail down several basic properties of the universe and provided overwhelming evidence that dark energy is real.

But Seife writes that proving dark energy's existence "beyond a reasonable doubt" fell to the Sloan Digital Sky Survey, which provided high-precision locations for a quarter-million galaxies. Superposing the locations on the WMAP temperature variation image revealed boosts in CMB energy at

*galaxy cluster locations, as expected if the
gravitational dimple caused by the clusters
had been flattened and stretched somewhat
by the antigravitational dark energy. —RA*

"Illuminating the Dark Universe"
by Charles Seife
Science, December 19, 2003

*Portraits of the earliest universe and the lacy pattern of
galaxies in today's sky confirm that the universe is
made up largely of mysterious dark energy and dark
matter. They also give the universe a firm age and a precise
speed of expansion.*

A lonely satellite spinning slowly through the void has
captured the very essence of the universe. In February,
the Wilkinson Microwave Anisotropy Probe (WMAP)
produced an image of the infant cosmos, of all of creation when it was less than 400,000 years old. The
brightly colored picture marks a turning point in the
field of cosmology: Along with a handful of other
observations revealed this year, it ends a decades-long
argument about the nature of the universe and confirms that our cosmos is much, much stranger than we
ever imagined.

Five years ago, *Science*'s cover sported the visage of
Albert Einstein looking shocked by 1998's Breakthrough
of the Year: the accelerating universe. Two teams of
astronomers had seen the faint imprint of a ghostly

force in the death rattles of dying stars. The apparent brightness of a certain type of supernova gave cosmologists a way to measure the expansion of the universe at different times in its history. The scientists were surprised to find that the universe was expanding ever faster, rather than decelerating, as general relativity—and common sense—had led astrophysicists to believe. This was the first sign of the mysterious "dark energy," an unknown force that counteracts the effects of gravity and flings galaxies away from each other.

Although the supernova data were compelling, many cosmologists hesitated to embrace the bizarre idea of dark energy. Teams of astronomers across the world rushed to test the existence of this irresistible force in independent ways. That quest ended this year. No longer are scientists trying to confirm the existence of dark energy; now they are trying to find out what it's made of, and what it tells us about the birth and evolution of the universe.

Lingering doubts about the existence of dark energy and the composition of the universe dissolved when the WMAP satellite took the most detailed picture ever of the cosmic microwave background (CMB). The CMB is the most ancient light in the universe, the radiation that streamed from the newborn universe when it was still a glowing ball of plasma. This faint microwave glow surrounds us like a distant wall of fire. The writing on the wall—tiny fluctuations in the temperature (and other properties) of the ancient light—reveals what the universe is made of.

Long before there were stars and galaxies, the universe was made of a hot, glowing plasma that roiled under the competing influences of gravity and light. The big bang had set the entire cosmos ringing like a bell, and pressure waves rattled through the plasma, compressing and expanding and compressing clouds of matter. Hot spots in the background radiation are the images of compressed, dense plasma in the cooling universe, and cold spots are the signature of rarefied regions of gas.

Just as the tone of a bell depends on its shape and the material it's made of, so does the "sound" of the early universe—the relative abundances and sizes of the hot and cold spots in the microwave background— depend on the composition of the universe and its shape. WMAP is the instrument that finally allowed scientists to hear the celestial music and figure out what sort of instrument our cosmos is.

The answer was disturbing and comforting at the same time. The WMAP data confirmed the incredibly strange picture of the universe that other observations had been painting. The universe is only 4% ordinary matter, the stuff of stars and trees and people. Twenty-three percent is exotic matter: dark mass that astrophysicists believe is made up of an as-yet-undetected particle. And the remainder, 73%, is dark energy.

The tone of the cosmic bell also reveals the age of the cosmos and the rate at which it is expanding, and WMAP has nearly perfect pitch. A year ago, a cosmologist would likely have said that the universe is between 12 billion and 15 billion years old. Now the estimate is 13.7 billion years, plus or minus a few hundred thousand. Similar calculations based on WMAP data have also pinned down the rate of the universe's expansion—71 kilometers per second per megaparsec, plus or minus a few hundredths—and the universe's "shape": slate flat. All the arguments of the last few decades about the basic properties of the universe—its age, its expansion rate, its composition, its density—have been settled in one fell swoop.

As important as WMAP is, it is not this year's only contribution to cosmologists' understanding of the history of the universe. The Sloan Digital Sky Survey (SDSS) is mapping out a million galaxies. By analyzing the distribution of those galaxies, the way they clump and spread out, scientists can figure out the forces that cause that clumping and spreading—be they the gravitational attraction of dark matter or the antigravity push of dark energy. In October, the SDSS team revealed its analysis of the first quarter-million galaxies it had collected. It came to the same conclusion that the WMAP researchers had reached: The universe is dominated by dark energy.

This year scientists got their most direct view of dark energy in action. In July, physicists superimposed the galaxy-clustering data of SDSS on the microwave data of WMAP and proved—beyond a reasonable doubt—that dark energy must exist. The proof relies on a phenomenon known as the integrated Sachs-Wolfe effect. The remnant microwave radiation acted as a backlight, shining through the gravitational dimples caused by the galaxy clusters that the SDSS spotted. Scientists saw a gentle crushing—apparent as a slight shift toward shorter wavelengths—of the microwaves shining near those gravitational pits. In an uncurved universe such as our own, this can happen only if there is some antigravitational force—a dark energy— stretching out the fabric of spacetime and flattening the dimples that galaxy clusters sit in.

Some of the work of cosmology can now turn to understanding the forces that shaped the universe when it was a fraction of a millisecond old. After the universe burst forth from a cosmic singularity, the fabric of the newborn universe expanded faster than the speed of light. This was the era of inflation, and that burst of growth—and its abrupt end after less than 10^{-30} seconds—shaped our present-day universe.

For decades, inflation provided few testable hypotheses. Now the exquisite precision of the WMAP data is finally allowing scientists to test inflation directly. Each current version of inflation proposes a slightly different scenario about the precise nature of the inflating force, and each makes a concrete prediction about the CMB, the distribution of galaxies, and even the clustering of gas clouds in the later universe. Scientists are just beginning to winnow out a handful of theories and test some make-or-break hypotheses. And as the SDSS data set grows—yielding information on distant quasars and gas clouds as well as the distribution of galaxies—scientists will challenge inflation theories with more boldness.

The properties of dark energy are also now coming under scrutiny. WMAP, SDSS, and a new set of supernova observations released this year are beginning to give scientists a handle on the way dark energy reacts to being stretched or squished. Physicists have already had to discard some of their assumptions about dark energy. Now they have to consider a form of dark energy that might cause all the matter in the universe to die a violent and sudden death. If the dark energy is stronger than a critical value, then it will eventually tear apart galaxies, solar systems, planets, and even atoms themselves in a "big rip." (Not to worry; cosmologists aren't losing sleep about the prospect.)

For the past 5 years, cosmologists have tested whether the baffling, counterintuitive model of a universe made of dark matter and blown apart by dark energy could be correct. This year, thanks to WMAP, the SDSS data, and new supernova observations, they know the answer is yes—and they're starting to ask new questions. It is, perhaps, a sign that scientists will finally begin to understand the beginning.

Reprinted with permission from Seife, Charles, "Illuminating the Dark Universe," SCIENCE 302:2038-2039 (2003). © 2003 AAAS.

The subtle temperature variations seen in the cosmic microwave background (CMB) provide the basic grist for modern cosmology theories.

The next article, by Wayne Hu and Martin White, describes the acoustic imprint on the CMB that conveys a great deal of information about the early universe and supports the theory of inflation.

The acoustics of everyday life is sound: conversation, music, traffic noise. But at its root, sound exists because the air can be compressed and decompressed by oscillating pressure waves.

Similarly, the early universe had "sound" when it was a gaseous plasma of charged particles and photons. Gravitational forces compressed the plasma until pressure from photons resisted and reversed the motion, leading to alternating compressions (hot spots) and decompressions (cold spots).

When the universe had cooled enough to allow protons to capture electrons, about 380,000 years after the big bang, charged particles no longer scattered photons, and the sound froze. The photons propagated away, taking with them their compression-resisting force, as well as the hot and cold acoustic pattern that evolved into the CMB through cosmic expansion.

Here, the authors show how the acoustic imprint provides information about the relative amount of mass and energy in the universe. —RA

"The Cosmic Symphony"
by Wayne Hu and Martin White
Scientific American, February 2004

New observations of the cosmic microwave background radiation show that the early universe resounded with harmonious oscillations

In the beginning, there was light. Under the intense conditions of the early universe, ionized matter gave off radiation that was trapped within it like light in a dense fog. But as the universe expanded and cooled, electrons and protons came together to form neutral atoms, and matter lost its ability to ensnare light. Today, some 14 billion years later, the photons from that great release of radiation form the cosmic microwave background (CMB).

Tune a television set between channels, and about 1 percent of the static you see on the screen is from the CMB. When astronomers scan the sky for these microwaves, they find that the signal looks almost identical in every direction. The ubiquity and constancy of the CMB is a sign that it comes from a simpler past, long before structures such as planets, stars and galaxies formed. Because of this simplicity, we can predict the properties of the CMB to exquisite accuracy. And in the past few years, cosmologists have been able to compare these predictions with increasingly precise observations from microwave telescopes carried by balloons and spacecraft. This research has brought us closer to answering some age-old questions: What is the universe made of? How old is it? And where did objects in the universe, including our planetary home, come from?

Arno Penzias and Robert Wilson of AT&T Bell Laboratories detected the CMB radiation in 1965 while trying to find the source of a mysterious background noise in their radio antenna. The discovery firmly established the big bang theory, which states that the early universe was a hot, dense plasma of charged particles and photons. Since that time, the CMB has been cooled by the expansion of the universe, and it is extremely cold today—comparable to the radiation released by a body at a temperature of 2.7 kelvins (that is, 2.7 degrees Celsius above absolute zero). But when the CMB was released, its temperature was nearly 3,000 kelvins (or about 2,727 degrees C).

In 1990 a satellite called COBE (for Cosmic Background Explorer) measured the spectrum of the CMB radiation, showing it to have exactly the expected form. Overshadowing this impressive achievement, however, was COBE's detection of slight variations—at the level of one part in 100,000—in the temperature of the CMB from place to place in the sky. Observers had been diligently searching for these variations for more than two decades because they hold the key to understanding the origin of structure in the universe: how the primordial plasma evolved into galaxies, stars and planets.

Since then, scientists have employed ever more sophisticated instruments to map the temperature variations of the CMB. The culmination of these efforts was the launch in 2001 of the Wilkinson Microwave

Anisotropy Probe (WMAP), which travels around the sun in an orbit 1.5 million kilometers beyond Earth's. The results from WMAP reveal that the CMB temperature variations follow a distinctive pattern predicted by cosmological theory: the hot and cold spots in the radiation fall into characteristic sizes. What is more, researchers have been able to use these data to precisely estimate the age, composition and geometry of the universe. The process is analogous to determining the construction of a musical instrument by carefully listening to its notes. But the cosmic symphony is produced by some very strange players and is accompanied by even stranger coincidences that cry out for explanation.

Our basic understanding of the physics behind these observations dates back to the late 1960s, when P. James E. Peebles of Princeton University and graduate student Jer Yu realized that the early universe would have contained sound waves. (At almost the same time, Yakov B. Zel'dovich and Rashid A. Sunyaev of the Moscow Institute of Applied Mathematics were coming to very similar conclusions.) When radiation was still trapped by matter, the tightly coupled system of photons, electrons and protons behaved as a single gas, with photons scattering off electrons like ricocheting bullets. As in the air, a small disturbance in gas density would have propagated as a sound wave, a train of slight compressions and rarefactions. The compressions heated the gas and the rarefactions cooled it, so any disturbance in the early universe resulted in a shifting pattern of temperature fluctuations.

Sounding Out Origins

When distances in the universe grew to one thousandth of their current size—about 380,000 years after the big bang—the temperature of the gas decreased enough for the protons to capture the electrons and become atoms. This transition, called recombination, changed the situation dramatically. The photons were no longer scattered by collisions with charged particles, so for the first time they traveled largely unimpeded through space. Photons released from hotter, denser areas were more energetic than photons emitted from rarefied regions, so the pattern of hot and cold spots induced by the sound waves was frozen into the CMB. At the same time, matter was freed of the radiation pressure that had resisted the contraction of dense clumps. Under the attractive influence of gravity, the denser areas coalesced into stars and galaxies. In fact, the one-in-100,000 variations observed in the CMB are of exactly the right amplitude to form the large-scale structures we see today.

Yet what was the prime mover, the source of the initial disturbances that triggered the sound waves? The question is troubling. Imagine yourself as an observer witnessing the big bang and the subsequent expansion. At any given point you will see only a finite region of the universe that encompasses the distance light has traveled since the big bang. Cosmologists call the edge of this region the horizon, the place beyond which you cannot see. This region continuously grows until it reaches the radius of the observable universe today. Because information cannot be conveyed faster than light, the horizon defines the sphere of influence of any physical mechanism. As we go backward in time to search for the origin of structures of a particular physical size, the horizon eventually becomes smaller than the structure. Therefore, no physical process that obeys causality can explain the structure's origin. In cosmology, this dilemma is known as the horizon problem.

Fortunately, the theory of inflation solves the horizon problem and also provides a physical mechanism for triggering the primordial sound waves and the seeds of all structure in the universe. The theory posits a new form of energy, carried by a field dubbed the "inflaton," which caused an accelerated expansion of the universe in the very first moments after the big bang. As a result, the observable universe we see today is only a small fraction of the observable universe before inflation. Furthermore, quantum fluctuations in the inflaton field, magnified by the rapid expansion, provide initial disturbances that are approximately equal on all scales—that is, the disturbances to small regions have the same magnitude as those affecting large regions. These disturbances become fluctuations in the energy density from place to place in the primordial plasma.

Evidence supporting the theory of inflation has now been found in the detailed pattern of sound waves in the CMB. Because inflation produced the density disturbances all at once in essentially the first moment of creation, the phases of all the sound waves were synchronized. The result was a sound spectrum with overtones much like a musical instrument's. Consider blowing into a pipe that is open at both ends. The fundamental frequency of the sound corresponds to a wave (also called a mode of vibration) with maximum air displacement at either end and minimum displacement in the middle. The wavelength of the fundamental mode is twice the length of the pipe. But the sound also has a series of overtones corresponding to wavelengths that are integer fractions of the fundamental wavelength: one half, one third, one fourth and so on. To put it another way, the frequencies of the overtones are two, three, four or more times as high as the fundamental frequency. Overtones are what distinguish a Stradivarius from an ordinary violin; they add richness to the sound.

The sound waves in the early universe are similar, except now we must imagine the waves oscillating in time instead of space. In this analogy, the length of the pipe represents the finite duration when sound waves traveled through the primordial plasma; the waves start at inflation and end at recombination about 380,000 years later. Assume that a certain region of space has a maximum positive displacement—that is, maximum temperature—at inflation. As the sound waves propagate, the density of the region will begin to oscillate, first heading toward average temperature (minimum displacement) and then toward minimum temperature (maximum negative displacement). The wave that causes the region to reach maximum negative displacement exactly at recombination is the fundamental wave of the early universe. The overtones have wavelengths that are integer fractions of the fundamental wavelength.

Oscillating two, three or more times as quickly as the fundamental wave, these overtones cause smaller regions of space to reach maximum displacement, either positive or negative, at recombination.

How do cosmologists deduce this pattern from the CMB? They plot the magnitude of the temperature variations against the sizes of the hot and cold spots in a graph called a power spectrum. The results show that the regions with the greatest variations subtend about one degree across the sky, or nearly twice the size of the full moon. (At the time of recombination, these regions had diameters of about one million light-years, but because of the 1,000-fold expansion of the universe since then, each region now stretches nearly one billion light-years across.) This first and highest peak in the power spectrum is evidence of the fundamental wave, which compressed and rarefied the regions of plasma to the maximum extent at the time of recombination. The subsequent peaks in the power spectrum represent the temperature variations caused by the overtones. The series of peaks strongly supports the theory that inflation triggered all the sound waves at the same time. If the perturbations had been continuously generated over time, the power spectrum would not be so harmoniously ordered. To return to our pipe analogy, consider the cacophony that would result from blowing into a pipe that has holes drilled randomly along its length.

The theory of inflation also predicts that the sound waves should have nearly the same amplitude on all scales. The power spectrum, however, shows a sharp drop-off in the magnitude of temperature variations after the third peak. This discrepancy can be explained by the fact that sound waves with short wavelengths dissipate. Because sound is carried by the collisions of particles in gas or plasma, a wave cannot propagate if its wavelength is shorter than the typical distance traveled by particles between collisions. In air, this distance is a negligible 10^{-5} centimeter. But in the primordial plasma just before recombination, a particle would typically travel some 10,000 light-years before striking another. (The universe at this stage was dense only in comparison with the modern universe, which is about a billion times as rarefied.) As measured today, after its 1,000-fold expansion, that scale is about 10 million light-years. Therefore, the amplitudes of the peaks in the power spectrum are damped below about 10 times this scale.

Just as musicians can distinguish a world-class violin from an ordinary one by the richness of its overtones, cosmologists can elucidate the shape and composition of the universe by examining the fundamental frequency of the primordial sound waves and the strength of the overtones. The CMB reveals the angular size of the most intense temperature variations—how large these hot and cold spots appear across the sky—which in turn tells us the frequency of the fundamental sound wave. Cosmologists can precisely estimate the actual size of this wave at the time of recombination because they know how quickly sound propagated in the primordial plasma. Likewise, researchers can determine the distance CMB photons have traveled before reaching Earth—about 45 billion light-years. (Although the photons have traveled for only about 14 billion years, the expansion of the universe has elongated their route.)

So cosmologists have complete information about the triangle formed by the wave and can check whether its angles add up to 180 degrees—the classic test of spatial curvature. They do so to high precision, showing that aside from the overall expansion, the universe obeys the laws of Euclidean geometry and must be very close to spatially flat. And because the geometry of the universe depends on its energy density, this finding implies that the average energy density is close to the so-called critical density—about 10^{-29} gram per cubic centimeter.

The next thing cosmologists would like to know is the exact breakdown of the universe's matter and energy. The amplitudes of the overtones provide the key. Whereas ordinary sound waves are driven solely by gas pressure, the sound waves in the early universe were modified by the force of gravity. Gravity compresses the gas in denser regions and, depending on the phase of the sound wave, can alternately enhance or counteract sonic compression and rarefaction. Analyzing the modulation of the waves reveals the strength of gravity, which in turn indicates the matter-energy composition of the medium.

As in today's universe, matter in the early universe fell into two main categories: baryons (protons and neutrons), which make up the bulk of so-called ordinary matter, and cold dark matter, which exerts gravity but has never been directly observed because it does not interact with ordinary matter or light in any noticeable way. Both ordinary matter and dark matter supply mass to the primordial gas and enhance the gravitational pull, but only ordinary matter undergoes the sonic compressions and rarefactions. At recombination, the fundamental wave is frozen in a phase where gravity enhances its compression of the denser regions of gas. But the first overtone, which has half the fundamental wavelength, is caught in the opposite phase—gravity is attempting to compress the plasma while gas pressure is trying to expand it. As a result, the temperature variations caused by this overtone will be less pronounced than those caused by the fundamental wave.

This effect explains why the second peak in the power spectrum is lower than the first. And by comparing the heights of the two peaks, cosmologists can gauge the relative strengths of gravity and radiation pressure in the early universe. This measurement indicates that baryons had about the same energy density as photons at the time of recombination and hence constitute about 5 percent of the critical density today. The result is in spectacular agreement with the number derived from studies of light-element synthesis by nuclear reactions in the infant universe.

The general theory of relativity, however, tells us that matter and energy gravitate alike. So did the gravity of the photons in the early universe also enhance the temperature variations? It did, in fact, but another effect counterbalanced it. After recombination, the CMB photons from denser regions lost more energy than photons from less dense areas, because they were climbing out of deeper gravitational-potential wells. This process, called the Sachs-Wolfe effect, reduced the amplitude of the temperature variations in the CMB, exactly negating the enhancement caused by the gravity of the photons. For regions of the early universe that were too big to undergo acoustic oscillations—that is, regions stretching more than one degree across the sky—temperature variations are solely the result of the Sachs-Wolfe effect. At these scales, paradoxically, hot spots in the CMB represent less dense regions of the universe.

Finally, cosmologists can use the CMB to measure the proportion of dark matter in the universe. The gravity from baryons alone could not have modulated the temperature variations much beyond the first peak in the power spectrum. An abundance of cold dark matter was needed to keep the gravitational- potential wells sufficiently deep. By measuring the ratios of the heights of the first three peaks, researchers have determined that the density of cold dark matter must be roughly five times the baryon density. Therefore, dark matter constitutes about 25 percent of the critical density today.

Remarkable Concord

Unfortunately, these calculations of the modern universe's matter and energy leave about 70 percent of the critical density unspecified. To make up the difference, theorists have posited a mysterious component called dark energy, whose relative influence has grown as the universe has expanded. We are thus led by degrees to an improbable conclusion: most of the universe today is composed of invisible dark matter and dark energy. Worse yet, dark matter and dark energy seem to be coincidentally comparable in energy density today, even though the former vastly outweighed the latter at recombination. Physicists dislike coincidences; they prefer to explain the world in terms of cause and effect rather than dumb luck. What is more, another mysterious component, the inflaton, dominated the very early universe and seeded cosmic structure. Why should we believe a cosmological model that is based on the seemingly fanciful introduction of three enigmatic entities?

One reason is that these three entities explain a wealth of previously known facts. Dark matter was first postulated in the 1930s to explain measurements of the local mass density in galaxy clusters. Albert Einstein introduced the concept of dark energy in 1917 when he included the so-called cosmological constant in his equations to counteract the influence of gravity. He later disavowed the constant, but it was resurrected in the 1990s, when observations of distant supernovae showed that the expansion of the universe is accelerating. The energy densities of dark matter and dark energy, as measured from the CMB, are in striking accord with these astronomical observations.

Second, the standard cosmological model has predictive power. In 1968 Joseph Silk (now at the University of Oxford) predicted that the small-scale acoustic peaks in the CMB should be damped in a specific, calculable way. As a result, the corresponding radiation should gain a small but precisely known polarization. (Polarized radiation is oriented in a particular direction.) One might assume that the CMB would be unpolarized because the scattering of the photons in the primordial plasma would have randomized their direction. But on the small scales where damping occurs, photons can travel with relatively few scatterings, so they retain directional information that is imprinted as a polarization of the CMB. This acoustic polarization was measured by the Degree Angular Scale Interferometer (an instrument operated at the Amundsen-Scott South Pole Station in Antarctica) and later by WMAP; the value was in beautiful agreement with predictions. WMAP also detected polarization on larger scales that was caused by scattering of CMB photons after recombination.

Furthermore, the existence of dark energy predicts additional phenomena in the CMB that are beginning to be observed. Because dark energy accelerates the expansion of the universe, it weakens the gravitational-potential wells associated with the clustering of galaxies. A photon traveling through such a region gets a boost in energy as it falls into the potential well, but because the well is shallower by the time the photon climbs back out, it loses less energy than it previously gained. This phenomenon, called the integrated Sachs-Wolfe effect, causes large-scale temperature variations in the CMB. Observers have recently seen hints of this correlation by comparing large structures in galaxy surveys with the WMAP data. The amount of dark energy needed to produce the large-scale temperature variations is consistent with the amount inferred from the acoustic peaks and the distant supernovae. As the data from the galaxy surveys improve and other tracers of the large-scale structure of the universe become available, the integrated Sachs-Wolfe effect could become an important source of information about dark energy.

No Requiem Yet

The CMB may also provide crucial new evidence that could explain what happened during the very first moments after the big bang. Few aspects of cosmology are more bizarre than the period of inflation. Did the universe really inflate, and, if so, what was the nature of the inflaton, the theoretical field that caused the rapid expansion? Current measurements of the CMB have dramatically strengthened the case for the simplest models of inflation, which assume that the amplitudes of the initial density fluctuations were the same at all scales. But if more detailed observations of the CMB reveal that the amplitudes varied at different scales, the simple inflation models would be in trouble. More baroque alternatives would need to be invoked or altogether different paradigms adopted.

Another exciting possibility is that we could learn about the physics of inflation by determining the energy scale at which it took place. For example, physicists believe that the weak nuclear force and the electromagnetic force were different aspects of a single electroweak force when the universe was hotter than 10^{15} kelvins. If researchers determine that inflation occurred at this energy scale, it would strongly imply that the inflaton had something to do with electroweak unification. Alternatively, inflation could have occurred at the much higher temperatures at which the electroweak force merges with the strong nuclear force. In this case, inflation would most likely be associated with the grand unification of the fundamental forces.

A distinctive signature in the CMB could allow researchers to settle this issue. In addition to spawning density perturbations, inflation created fluctuations in the fabric of spacetime itself. These fluctuations are gravitational waves whose wavelengths can stretch across the observable universe. The amplitude of these gravitational waves is proportional to the square of the energy scale at which inflation took place. If inflation occurred at the high energies associated with grand unification, the effects might be visible in the polarization of the CMB.

Last, further observations of the CMB could shed some light on the physical nature of dark energy. This entity might be a form of vacuum energy, as Einstein had hypothesized, but its value would have to be at least 60 and perhaps as much as 120 orders of magnitude as small as that predicted from particle physics. And why is dark energy comparable to dark matter in density now and apparently only now? To answer these questions, researchers can take advantage of the fact that CMB photons illuminate structures across the entire observable universe. By showing the amplitude of density fluctuations at different points in cosmic history, the CMB can reveal the tug-of war between matter and dark energy.

Measurements of two CMB phenomena could be particularly useful. The first, called the Sunyaev-Zel'dovich effect, occurs when CMB photons are scattered by the hot ionized gas in galaxy clusters. This effect allows galaxy clusters to be identified during the crucial period, about five billion years ago, when dark energy began to accelerate the expansion of the universe. The number of galaxy clusters, in turn, indicates the amplitude of density fluctuations during this time. The second phenomenon, gravitational lensing, happens when CMB photons pass by a particularly massive structure that bends their trajectories and hence distorts the pattern of temperature and polarization variations. The degree of lensing reveals the amplitude of the mass density fluctuations associated with these structures.

To conduct these investigations of inflation and dark energy, however, researchers will need a new generation of CMB telescopes that can observe the radiation with even greater sensitivity and resolution. In 2007 the European Space Agency plans to launch the Planck spacecraft, a microwave observatory that will be placed in the same orbit as WMAP. Planck will be able to measure CMB temperature differences as small as five millionths of a kelvin and detect hot and cold spots that subtend less than a tenth of a degree across the sky. Such measurements will enable scientists to glimpse the full range of acoustic oscillations in the CMB and thus sharpen their picture of the inflationary spectrum. A multitude of ground-based experiments are also under way to study CMB effects associated with structure in the current epoch of accelerated expansion.

Although the standard cosmological model appears to work remarkably well as a phenomenological description of the universe, a deeper understanding of its mysteries awaits the findings of these experiments. It seems clear that the cosmic symphony will continue to enchant its listeners for some time to come.

The study of X-ray emissions from vast clusters of galaxies described in the next article adds further support to the view that cosmic expansion is accelerating.

The research, conducted in 2004, complements the results found from measurements of type Ia supernovas and the cosmic microwave background, which are described earlier in this chapter.

The X-ray measurements are from the Chandra X-ray Observatory, part of NASA's fleet of great observatories which includes the Hubble Space Telescope. Chandra is located in Cambridge, Massachusetts, as part of the Harvard-Smithsonian Center for Astrophysics.

The Chandra team's approach resembles, in many ways, the one used in the type Ia supernova study, which estimated distances from supernova brightness and speeds from the redshift of the supernova light to estimate a history of cosmic expansion.

In the Chandra approach, the researchers analyzed the brightness and energy spectra of X-ray emissions from twenty-six clusters to find the distances and combined these with the observations of others on how fast the clusters were receding.

The researchers targeted clusters in a range of distances between 1 billion and 8 billion light-years, in an effort to pinpoint the time when the acceleration began to increase. They found it happened about 6 billion years ago, in agreement with more recent supernova studies. —RA

"Galaxy Clusters Bear Witness to Universal Speed-Up"
by Adrian Cho
Science, May 21, 2004

No matter how you measure it, the expansion of the universe is speeding up. Or so say researchers who have studied the x-rays coming from far-flung clusters of galaxies. The new results bolster previous measurements of individual exploding stars called supernovae and the indirect analysis of all-pervading radiation known as the cosmic microwave background. "This adds a third pillar to the evidence supporting the accelerating universe," says Michael Turner, a cosmologist at the University of Chicago. But the new results depend on an assumption that skeptics may question.

In 1998, two independent research teams reached the stunning conclusion that some distant exploding stars were farther away than they would be if expansion were constant or slowing down as expected (*Science*, 27 February 1998, p. 1298). The case for accelerating expansion gained strength last year. Researchers working with the Wilkinson Microwave Anisotropy Probe satellite measured tiny spatial variations in the cosmic microwave background and deduced that the universe must consist mainly of space-stretching "dark energy" (*Science*, 14 February 2003, p. 991). But the analysis did not directly measure expansion.

Now, x-rays from clusters of thousands of galaxies provide more direct evidence of accelerated expansion, report researchers working with NASA's orbiting Chandra X-ray Observatory. Led by astronomer Steven Allen of the University of Cambridge, U.K., the team used the energy spectra of the x-rays from 26 clusters to infer the amount of hot gas in each cluster. That in turn determined the amount of x-rays the cluster emits. The researchers deduced the distance to a cluster by noting how bright it appeared in x-rays, just as a driver might judge the distance to another car by noting the brightness of its taillights.

The researchers then compared the results with how far away each cluster would be if expansion were constant, as revealed by the redshift of its light. The comparisons confirmed that the universe began accelerating 6 billion years ago, they reported on 18 May at a press conference at NASA Headquarters in Washington, D.C.

The analysis assumes that hot gas accounts for the same fraction of matter in every cluster, no matter how old. That isn't obviously true, says astrophysicist Neta Bahcall of Princeton University in New Jersey, especially for the oldest, most distant clusters. Alastair Edge, an astrophysicist at the University of Durham, U.K., agrees but says most researchers will be persuaded in the end. "Everyone's going to scratch their heads for a few months," he says, "and then they'll probably accept it."

Reprinted with permission from Cho, Adrian, "Galaxy Clusters Bear Witness to Universal Speed-Up," SCIENCE 304:1092 (2004). © 2004 AAAS.

We understand very little about dark energy besides its apparent existence, but efforts described in the next article are under way to change this.

A basic question about the energy is whether it has changed over time. Future supernova surveys of the sort that first uncovered dark energy will answer this by taking many more samples, including some from the universe's early history.

Several experiments are planned that will survey galaxy clusters to constrain dark energy's strength, which affects cluster abundance over time. The clusters consist of thousands of galaxies gravitationally bound into huge structures.

Similarly, dark energy's influence on the universe's mass distribution is expected to be perceived from surveys of weak gravitational lensing, which is the moderate bending of light from distant galaxies by intervening galaxies.

More down-to-earth efforts include looking at neutrinos. The energy scales of dark energy and the masses of the three known neutrinos are about the same, suggesting a quantum mechanical connection. Some models that seem promising require a fourth neutrino, which is currently being sought.

On the theoretical side, some investigators are tweaking Einstein's general theory of relativity to accommodate dark energy. —RA

"Dark Doings"
by Ron Cowen
Science News, **May 22, 2004**

Searching for signs of a force that may be everywhere . . . or nowhere

Ever since 1998, Robert Caldwell has been obsessed with something dark and repulsive. He spends nearly every waking moment trying to comprehend a mysterious entity that may be undermining gravity and pulling everything apart, making the universe expand at a faster and faster rate. This presumed force, sometimes called dark energy, might ultimately rip apart every object in the cosmos, from the tiniest of atoms to gargantuan clusters of galaxies (SN: 2/28/04, p. 132: Available to subscribers at http://www. sciencenews.org/articles/20040228/ob3.asp). "It's both fascinating and terrifying," says Caldwell, a cosmologist at Dartmouth College in Hanover, N.H.

Caldwell has partners in his obsession, among them other theorists and the astronomers who dropped the bombshell about cosmic acceleration onto the scientific community 6 years ago. That's when two studies of distant exploding stars first revealed that the universe is accelerating its rate of expansion—exactly the opposite of what had been expected. The mutual gravity of all the matter in the cosmos ought to be slowing down the expansion that began with the Big Bang. The new observations led the teams to propose that there was something previously unimagined pushing everything away from everything else.

"It's kind of amazing that it's only been 6 years" since those findings, says observer John Tonry of the University of Hawaii in Honolulu, "because it now seems so much a part of the canonical lore that we believe about the universe."

Now, several teams of researchers are conducting experiments and planning ambitious new ones to investigate this suspected force. On the cosmic scale, astronomers are developing new sky surveys in search of supernovas while also studying the shapes of galaxies and the evolution of galaxy clusters. On a small scale, particle physicists are turning to atom-smashing experiments that may reveal whether the mystery lies in hidden spatial dimensions or in as-yet-undiscovered fundamental particles.

"When you have a big problem, you throw everything you can at it," says Joseph Lykken of Fermi National Accelerator Laboratory in Batavia, Ill.

Cosmic acceleration is "not just another mystery," says Lykken. "It's getting at something fundamental in our understanding of gravity, energy, and quantum theory. It may take us 20 years to [figure it out], but it will open a whole new chapter in physics, a revolution in our understanding of the world."

Seeking Supernovas

In trying to identify the origin of this cosmic push, researchers are simultaneously considering two divergent and equally bizarre scenarios. In one approach, researchers embrace the concept of dark energy and look for the fingerprints of this unseen entity, which may spread uniformly through the cosmos and be an intrinsic property of empty space.

The other strategy denies dark energy's existence and instead seeks to explain cosmic acceleration by modifying the laws of gravity. According to this perspective, the wildly successful theory of gravity developed by Albert Einstein needs revision, especially as it describes gravity over large distances.

If dark energy is real, a special type of supernova may shine a light on its properties. Known as type 1a supernovas, these stellar explosions all have about the same intrinsic brightness, like light bulbs of similar wattage. A comparison of that fixed brightness to the brightness with which each supernova appears on the sky enables astronomers to measure the distance to each of these stellar explosions. By recording the spectrum of light emitted by a type 1a supernova, astronomers learn how fast its host galaxy was receding at the time the supernova erupted. With the information on the distance and recession velocities from many supernovas, astronomers can reconstruct how fast the universe was expanding at different times during its history.

From the 200 or so type 1a supernovas that astronomers have now studied, they've deduced that galaxies today are flying apart faster than they did 5 billion years ago—prima facie evidence for runaway expansion. Now, researchers want to learn whether the presumed dark energy has had the same density throughout cosmic history.

If its density is constant, then dark energy may resemble what Einstein called the cosmological constant—an unchanging property of empty space that imbues the universe with a constant acceleration. If dark energy's density varies, it could either increase in strength and rip the universe apart, or it could fade away. In the latter case, the gravitational tug of all the matter in the universe would eventually cause the cosmos to collapse.

By the end of the decade, astronomers hope to have a telescope that will find thousands of type 1a supernovas and produce enough data to reveal whether or not dark energy has varied. For example, the Supernova Acceleration Probe, an orbiting satellite bearing a

1.8-meter telescope and the largest solid-state camera ever constructed, would image and take spectra of some 6,000 supernovas. If it gets funded by NASA and the Department of Energy, the 3-year mission could be launched by 2010. However, the project's funding has been delayed by NASA's recent presidential mandate to focus on human exploration of the moon and Mars.

Cluster Connection

In addition to studying explosions inside individual galaxies, astronomers are also trying to glimpse dark energy's effects by determining when and how clusters of galaxies coalesced.

The evolution of clusters—or any massive cosmic structure whose formation depends on gravitational attraction—is closely tied to the strength of dark energy. Early in the universe, when the density of matter was high, gravitational attraction would have handily won the tug-of-war with dark energy's repulsive force. Later, as the universe expanded more and more, matter became more dilute, permitting dark energy's push to overpower it. So, in a universe brimming with dark energy, clusters must form early or they won't form at all.

The earlier the galaxy clusters formed, the stronger dark energy must be. To determine how far back in time most clusters coalesced, astronomers must find the most distant ones. One technique is to look for signs of the hot, X-ray-emitting gas that bathes clusters. The proposed Dark Universe Observatory, a suite of seven Earth-orbiting telescopes, would scan a large chunk of the sky in search of the X rays. Astronomers will then combine the X-ray data with information already in hand from the Sloan Digital Sky Survey, which has recorded the distances to several hundred thousand galaxies. The results are expected to indicate when galaxy clusters formed.

Other cluster watchers examine the cosmic microwave background, the radiation left over from the Big Bang. When photons from that background strike the hot gas surrounding a cluster, they gain energy. It's this shift in photon energy, known as the Sunyaev-Zeldovich effect, that John E. Carlstrom of the University of Chicago and his colleagues will be examining in unprecedented detail beginning late this summer.

Using their new Sunyaev-Zeldovich Array of six 3.5-m radio receivers at the Owens Valley Radio Observatory near Big Pine, Calif., Carlstrom and his collaborators expect to find thousands of new clusters. Several other teams are building similar radio telescopes. And in 2007, Carlstrom expects to have finished building an even more sensitive detector of clusters, a radio telescope at the South Pole.

Another search strategy for signs of dark energy takes advantage of a cosmic distortion known as gravitational lensing. Because any massive object causes space-time to curve, it can bend the path of a light ray emitted by a body, such as a galaxy, that lies behind it. The shape of that body appears distorted, as if the light had passed through a thick glass lens. In so-called weak lensing, light emitted by the outer parts of distant galaxies is distorted by the gravity of all the individual galaxies that lie in front of it.

Weak lensing relates to dark energy because the expansion rate of the universe determines how much volume lies between distant galaxies and Earth. Dark energy's push would increase the volume of space, making it more likely that light traveling to Earth from a distant galaxy would pass near other bodies and exhibit weak lensing. Dark energy would also require clumps of matter to begin coalescing into galaxies earlier in the history of the universe, also increasing the chances for lensing to occur.

To perceive the small effect of weak lensing, astronomers will have to study millions of galaxies distributed across the sky. The proposed Large Synoptic Survey Telescope, an 8-m ground-based instrument, could open for business in 2011. The orbiting Supernova Acceleration Probe could also lend a hand in weak-lensing studies.

Getting Particular

Dark energy may also reveal itself on the subatomic scale. Particle physicists at Fermilab and other high-energy physics laboratories are paying close attention to the neutrino, an elementary particle known to come in three flavors—tau, muon, and electron. A decade ago, scientists discovered that each type of neutrino could transform into the others. These so-called oscillations indicate that neutrinos, which for decades were thought to be massless, actually have some weight.

Theorists have homed in on what may be a deep connection between dark energy and particle physics. Mass and energy are equivalent, according to

Einstein, and scientists have noticed that the energy
scale associated with dark energy, about one-thousandth
of an electronvolt, is approximately the same as the
masses associated with the three known types of neutrinos. Assuming that this isn't a coincidence,
scientists have been trying to identify a single quantum mechanical description that applies to both dark
energy and neutrinos.

They've found that neutrino oscillations can be
described by a time-varying field that resembles the time-
varying dark energy described in some models. However,
these models also predict a fourth type of neutrino for
which there is yet no experimental evidence.

In a series of ongoing experiments at Fermilab, scientists are searching for that missing neutrino. In these
studies, a beam of muon neutrinos crosses a 30-foot
tank of mineral oil. Some 520 light detectors lining the
tank record the flashes that occur when neutrinos strike
carbon nuclei in the tank. An analysis of those flashes
has yet to reveal evidence of a fourth neutrino, but the
experiment is set to continue for another 2 years.

Gravity on the Fly

Despite the interest in time-varying dark energy, an
unchanging energy density akin to the cosmological
constant now appears to be the more accurate model for
the brand of dark energy that might exist in our universe, says cosmologist Sean Carroll of the University of
Chicago. Yet for Carroll and other theorists, the notion
of a cosmological constant is downright distasteful.

For starters, the only source that scientists have come up with for an unvarying dark-energy density is the energy associated with the vacuum of space. As described by quantum theory, the vacuum seethes with the relentless creation and annihilation of subatomic particles and their antiparticles. But calculations show that the density of this vacuum energy is a whopping 10^{120} times as big as that of dark energy. Such a glaring discrepancy makes it hard for Carroll and others to fully embrace the cosmological-constant model.

Then there's the cosmic-coincidence scandal. The density of matter in the universe has steadily declined since the Big Bang, and measurements show that today it's about the same as the density of dark energy predicted by the cosmological-constant models. There's only a 1 percent chance, Carroll calculates, that observers would be living at a time when the density of dark energy and matter were comparable. For some physicists, this match is too unlikely to be true.

Instead of accepting dark energy, these scientists would rather try to account for the acceleration of the universe's expansion by tinkering with Einstein's general theory of relativity.

Gia Dvali of New York University and his colleagues propose that gravity parts company with Einstein's theory because some of it leaks away into extra, hidden dimensions. They suggest that the universe as we know it—galaxies, stars, and familiar elementary particles—is confined to a four-dimensional space-time, called a brane, that's embedded in a higher-dimensional world.

Because gravity is an intrinsic property of all of space-time, however, it may be the only component of the cosmos that isn't trapped on this four-dimensional brane, Dvali suggests. He compares the scenario to what happens when a metal plate submerged in water is struck with a hammer. As the plate—representing the brane vibrates, some of the sound waves escape into the surrounding water—representing higher dimensions.

When gravitons, the particles that mediate gravitation attraction, escape the local brane, the gravitational force that remains within the brane diminishes. The weakening of gravity shows up as an increase in the rate of cosmic expansion. In this way, leaky gravity looks and behaves much like dark energy.

Dvali and his collaborators are still fleshing out their model, but it already has some concrete predictions measurable within our own solar system. Leaky gravity, it turns out, should cause the moon to tilt ever so slightly in its orbit about Earth. New measurements are looking for such a precession by using ground-based lasers that bounce off mirrors that the Apollo 11 astronauts left on the moon 3 decades ago.

Hidden dimensions and leaky gravity may also reveal themselves in experiments at Fermilab (SN: 2/19/00, p. 122: http://www.sciencenews.org/articles/20000219/bob9.asp). At extremely high energies, collisions between two particles, such as a proton and an antiproton, should produce a graviton, along with a spray of other particles. Those other particles will remain trapped on a four-dimensional brane, but the graviton can escape.

If it does so, then there ought to be a noticeable deficit in the amount of energy recorded. Such missing energy would serve as a signpost of the universe's higher dimensions and a gravitational theory that goes beyond that of Einstein.

No such missing energy has yet been detected, but physicists continue to search. "We had the capability to look for this before, but people didn't think [the notion of higher dimensions] was a reasonable idea. Now, people have started to take this seriously," says Lykken.

Whether it's dark energy that rules the universe or a kind of gravity that goes beyond what Einstein had imagined remains to be seen. Whatever the answer is, Caldwell notes, it's bound "to answer some deep questions about the universe."

3 Dark Matter

So far, much of this anthology has been concerned with the recent discovery of accelerating cosmic expansion and its implications for a mysterious and previously unknown dark energy that appears to make up some 70 percent of the universe.

An equally mysterious substance, dark matter, is thought to exist, and it also plays an important role in the evolution and fate of the universe. It outnumbers ordinary matter that we can see (such as stars and interstellar gas) by a factor of five to one or so, and it comprises about 25 percent of the universe.

But it is far from a newcomer to astronomers. It was first proposed in 1933 by Caltech's Fritz Zwicky to explain the galaxy orbits in the Coma cluster. He was only able to make sense of the observations by proposing that about 90 percent of the cluster mass must be undetectable except through its gravitational influence, a result found over and over in subsequent measurements of many other clusters.

In this article, David Cline summarizes current candidates for dark matter, which he says is almost certainly "a hitherto undiscovered type of elementary particle," as well as current and future efforts to snare it with earthbound detectors. —RA

"The Search for Dark Matter"
by David B. Cline
Scientific American, March 2003

Dark matter is usually thought of as something "out there." But we will never truly understand it unless we can bring it down to earth

The universe around us is not what it appears to be. The stars make up less than 1 percent of its mass; all the loose gas and other forms of ordinary matter, less than 5 percent. The motions of this visible material reveal that it is mere flotsam on an unseen sea of unknown material. We know little about that sea. The terms we use to describe its components, "dark matter" and "dark energy," serve mainly as expressions of our ignorance.

For 70 years, astronomers have steadily gathered circumstantial evidence for the existence of dark matter, and nearly everyone accepts that it is real. But circumstantial evidence is unsatisfying. It cannot conclusively rule out alternatives, such as modified laws of physics [see "Does Dark Matter Really Exist?" by Mordehai Milgrom; SCIENTIFIC AMERICAN, August 2002]. Nor does it reveal much about the properties of the supposed material. Essentially, all we know is that dark matter clumps together, providing a gravitational anchor for galaxies and larger structures such as galaxy clusters. It almost certainly consists of a hitherto undiscovered type of elementary particle. Dark energy, despite its confusingly similar name, is a separate substance that entered the picture only in 1998. It is spread uniformly through space, exerts a negative pressure and causes the expansion of the universe to accelerate.

Ultimately the details of these dark components will have to be filled in not by astronomy but by particle physics. Over the past eight years the two disciplines have pooled their resources, coming together at meetings such as the Symposia on Sources and Detection of Dark Matter and Dark Energy in the Universe. The next symposium will be held in February 2004 in Marina del Rey, Calif. The goal has been to find ways to detect and study dark matter using the same techniques that have been so successful for analyzing particles such as positrons and neutrinos. Rather than inferring its presence by looking at distant objects, scientists would seek the dark matter here on Earth.

The search for dark matter particles is among the most difficult experiments ever attempted in physics. (The search for particles of dark energy is even less tractable and has been put aside, at least for the time being.) At the first symposium, in February 1994, participants expressed a nearly total lack of confidence that a particle detector in an Earth-based lab could ever register dark matter. The sensitivity of even the best instruments was a factor of 1,000 too low to pick up hypothesized types of dark particles. But since then, detector sensitivity has improved 1,000-fold, and instrument builders expect soon to wring out another factor of 1,000. More than 15 years of research and development on detector methods are finally bearing fruit. We may soon know what the universe is really like. Either dark matter will prove to be real, or else the theories that underlie modern physics will have to fall on their swords.

Through the Looking Glass

What kind of particle could dark matter be made of? Astronomical observation and theory provide some general clues. It cannot be protons, neutrons, or anything that was once made of protons or neutrons, such as massive stars that became black holes. According to calculations of particle synthesis during the big bang, such particles are simply too few in number to make up the dark matter. Those calculations have been corroborated by measurements of primordial hydrogen, helium and lithium in the universe.

Nor can more than a small fraction of the dark matter be neutrinos, a lightweight breed of particle that zips through space and is unattached to any atom. Neutrinos were once a prominent possibility for dark matter, and their role remains a matter of discussion, but experiments have found that they are probably too lightweight [see "Detecting Massive Neutrinos," by Edward Kearns, Takaaki Kajita and Yoji Totsuka; SCIENTIFIC AMERICAN, August 1999]. Moreover, they are "hot"—that is, in the early universe they were moving at a velocity comparable to the velocity of light. Hot particles were too fleet-footed to settle into observed cosmic structures.

The best fit to the astronomical observations involves "cold" dark matter, a term that refers to some undiscovered particle that, when it formed, moved sluggishly. Although cold dark matter has its own problems in explaining cosmic structures [see "The Life Cycle of Galaxies," by Guinevere Kauffmann and Frank van den Bosch; SCIENTIFIC AMERICAN, June 2002], most cosmologists consider these problems minor compared with the difficulties faced by alternative hypotheses. The current Standard Model of elementary particles contains no examples of particles that could serve as cold dark matter, but extensions of the Standard Model—developed for reasons quite separate from the needs of astronomy—offer many plausible candidates.

By far the most studied extension of this kind is supersymmetry, so I will concentrate on this theory. Supersymmetry is an attractive explanation for dark matter because it postulates a whole new family of particles—one "superpartner" for every known elementary particle. These new particles are all heavier (hence more sluggish) than known particles. Several are natural candidates for cold dark matter. The one that gets the most attention is the neutralino, which is an amalgam of the superpartners of the photon (which transmits the electromagnetic force), the Z boson (which transmits the so-called weak nuclear force) and perhaps other particle types. The name is somewhat unfortunate: "neutralino" sounds much like "neutrino," and the two particles indeed share various properties, but they are otherwise quite distinct.

Although the neutralino is heavy by normal standards, it is generally thought to be the lightest supersymmetric particle. If so, it has to be stable: if a superparticle is unstable, it must decay into two lighter superparticles, and the neutralino is already the lightest. As the name implies, the neutralino has zero charge, so it is unaffected by electromagnetic forces (such as those involving light). The hypothesized mass, stability and neutrality of the neutralino satisfy all the requirements of cold dark matter.

The big bang theory gives an estimate of the number of neutralinos that were created within the hot primordial plasma of the cosmos. The plasma was a chaotic soup of all types of particles. No individual particle survived for long. It would quickly collide with another particle, annihilating both but producing new particles in the process; those new particles soon collided with others, in a cycle of destruction and creation. But as the universe cooled down and thinned out, the collisions became less violent, and the process ground to a halt. Particles condensed out one by one, beginning with those that tended to collide less often and proceeding to more collision-prone types.

Shy but No Hermit

The neutralino is a particularly collision-shy particle, so it froze out early on. At the time, the density of the universe was still very high, so a huge number of neutralinos were produced. In fact, based on the expected neutralino mass and its low tendency to collide, the total mass in

137

neutralinos almost exactly matches the inferred mass of dark matter in the universe. This correspondence is a strong sign that neutralinos are indeed dark matter.

To detect dark matter, scientists need to know how it interacts with normal matter. Astronomers assume that it interacts only by means of gravitation, the weakest of all the known forces of nature. If that is really the case, physicists have no hope of ever detecting it. But the astronomers' assumption is probably just a convenient approximation—something that lets them describe cosmic structures without worrying about the detailed properties of the particles.

Theories of supersymmetry predict that the neutralino will interact by a force stronger than gravitation: the weak nuclear force. This is similar to the interaction that betrays neutrinos [see "The Search for Intermediate Vector Bosons," by David B. Cline, Carlo Rubbia and Simon van der Meer; SCIENTIFIC AMERICAN, March 1982]. The vast majority of neutralinos will slip through a slab of matter without interacting, but the occasional neutralino will hit an atomic nucleus. The unlucky particle will transfer a small amount of its energy to the nucleus.

The improbability and feebleness of the interaction are offset by the sheer number of particles. After all, dark matter is thought to dominate the galaxy. Being dark, it was never able to lose energy by emitting radiation, so it never could agglomerate into subgalactic clumps such as stars and planets. Instead it continues to suffuse interstellar space like a gas. Our solar system is orbiting around the center of the galaxy at 220 kilometers a second, so we are pushing through this gas at quite a clip.

Researchers estimate that a billion dark matter particles flow through every square meter every second.

Leszek Roszkowski and his team at the University of Lancaster in England recently carried out a complete calculation of the rates of neutralino interactions with normal matter. The rates are usually expressed as the number of events that would occur in a day in a single kilogram of normal matter. Depending on the theoretical details, the figures vary from 0.0001 to 0.1 event per kilogram a day. Current experiments are able to detect event rates in the high end of this range.

The main difficulty is no longer detector sensitivity but detector impurity. All materials on Earth, including the metal out of which the detectors are built, contain a trace amount of radioactive material such as uranium and thorium. The decay of this material produces particles that register much as dark matter would. Terrestrial radioactivity typically outpowers the putative neutralino signal by a factor of 10^6. If the detectors are located aboveground, cosmic rays worsen the situation by an equal factor. To identify dark matter particles with any confidence, researchers must reduce both these unwanted backgrounds a millionfold.

Turning the Other Cheek

Physicists thus face two challenges: to detect the inherently weak interaction of dark matter with ordinary matter and to screen out confounding noise. To take the first challenge first, several properties of matter can be used to record the recoil of a nucleus that has been struck by a neutralino. Perhaps the simplest of all possible methods is just to look for the heating that will occur when the recoiling nucleus plows into the surrounding matter and gives up its kinetic energy, thereby raising the temperature of the material slightly. To detect this heating, the material must be at a very low temperature to start with. This is the principle of a cryogenic detector.

Cryogenic detectors such as those used by two leading search programs, the Cryogenic Dark Matter Search (CDMS) and Edelweiss, are designed to measure individual phonons, or quanta of heat, in a material. They operate at a temperature of about 25 millikelvins and use thermistors to record the temperature rise in the various parts of the apparatus. Individual detectors have a mass of a few hundred grams, and researchers can stack a large number of detectors to reach a total mass of a few kilograms or more, thereby boosting the signal. The latest incarnation of CDMS, located inside the Soudan Mine in Minnesota, is scheduled to start taking data later this year.

A second method watches for another effect of the recoiling nucleus: ionization. The nucleus knocks some electrons off surrounding atoms, resulting in excited ions known as excimers. Those ions eventually recapture an electron and return to normal. In some materials, mainly noble gas liquids such as xenon, the process triggers the emission of light, called scintillation light. This is how excimer lasers—those used in eye surgery—work. For liquid xenon, the light is very intense and lasts about 10 nanoseconds. A photomultiplier can amplify the signal to detectable levels.

In the early 1990s the ZEPLIN project—led by HanGuo Wang and me at U.C.L.A. and Pio Picchi of the University of Turin in Italy—developed two-phase liquid-xenon detectors. These instruments amplify the light by introducing a layer of gas threaded by an electric field; the field accelerates the electrons that get kicked off by recoiling nuclei, thereby turning a handful of particles into an avalanche. Eventually it should be possible to construct a 10-metric-ton liquid-xenon detector, which should be sensitive to the neutralinos even if their interactivity is very low.

The xenon need not be in liquid form. Some detectors use it in gaseous form. Although the gas has a lower density than the liquid does, gas more readily reveals the trail left by the recoiling nucleus. The trail points back to the direction of the incoming dark matter, allowing a further check that a galactic neutralino is responsible. Detectors of this type are being developed for the Boulby underground laboratories in England.

Xenon is convenient because it has no natural long-lived radioactive isotopes (thus reducing the background noise) and is readily available in the atmosphere (after purification to remove radioactive krypton left over from nuclear bomb tests). But it is not the only material that scintillates. DAMA, an experiment being conducted at the Gran Sasso Laboratory near Rome, uses sodium iodide. With a mass of 100 kilograms, DAMA is the largest detector in the world.

Telling the Difference

Three steps are generally taken to cope with the other great challenge, overcoming the background noise from natural radioactivity and cosmic rays. First, researchers screen out cosmic rays by placing detectors deep underground and enclosing them in special shields. Second, they purify the detector material to reduce radioactive contamination. Third, they build special instruments to look for the telltale signs that distinguish dark matter from other particles.

Even when the first two steps are taken, they are not enough. Therefore, new dark matter detectors all take the third step, employing some form of event discrimination. The first line of defense is to look for an annual variation of the signal. The flux of dark matter should be higher in the northern summer, when Earth's orbital motion adds to the overall motion of the solar system through the galaxy, than in the northern winter, when Earth's motion subtracts from that of the solar system. The signal variation could be as high as a few percent.

The most advanced projects add a secondary detector, built using a different technology from that of the primary. The two detectors will respond to different types of particles in slightly different ways. For example, background particles tend to produce more ionization than a nucleus recoiling from a neutralino collision. By combining two detectors, this difference can be caught.

Using one or more of the above techniques, searches for dark matter signals started in earnest in the late

1980s. All but one have been null to date, which is not surprising, because they have only recently achieved the requisite sensitivity and noise tolerance. The lone exception is DAMA. Four years ago this project reported an observation of annual variation, which created excitement and skepticism in equal measure [see "Revenge of the WIMPs," by George Musser; News & Analysis, SCIENTIFIC AMERICAN, March 1999]. The problem was that DAMA does not use multiple detectors to discriminate between signal and noise. Three other experiments that do use multiple detectors have since cast doubt on DAMA's claims. Edelweiss, ZEPLIN I and CDMS I observed nothing in much of the range of parameters that DAMA had probed. The CDMS I team claimed a confidence level of 98 percent for the null result. If independent projects continue to come up empty-handed, the DAMA researchers will have to attribute their signal to radioactive processes or other noise.

The new generation of detectors should be able to rule neutralinos conclusively in or out. If they do not find anything, then supersymmetry must not be the solution that nature has chosen for the dark matter problem. Theorists would have to turn to other ideas, however distasteful that may now seem. But if the detectors do register and verify a signal, it would go down as one of the great accomplishments of the 21st century. The discovery of 25 percent of the universe (leaving only the dark energy unexplained) would obviously be the most spectacular implication. Other valuable information would follow. If detectors can spot particles of dark matter, particle accelerators such as CERN's Large

Hadron Collider near Geneva might be able to re-create them and conduct controlled experiments. The confirmation of supersymmetry would imply a vast number of new particles waiting to be discovered and would lend support to string theory, in which supersymmetry plays an integral role. The greatest mystery in modern astrophysics may soon be solved.

While we don't know exactly what dark matter is, we generally understand what it does—it binds galaxies together, bends light, molds intergalactic structures, and slows cosmic expansion.

In the next article, Ken Freeman summarizes recent developments in our understanding of dark matter at galactic scales, including advances in theory and observations that are helping to resolve some "important discrepancies."

Current models suggest galactic dark matter lies in dark halos enveloping the luminous parts of the parent galaxy. These halos extend much farther than the visible galaxy and considerably outweigh it. All these features are observed in our own large galaxy, the Milky Way.

But mass concentrations near halo cores don't always follow predictions, and the number of smaller satellite galaxies observed for the Milky

Way—twenty or so—is far fewer than the expected 500.

Freeman outlines several recent approaches for characterizing galaxy properties that address these problems. He focuses on surveys of weak gravitational lensing, or the moderate gravitational bending of light as it passes by an intervening galaxy.

Weak lensing, the moderate bending of light caused by a mass concentration, can reveal dark matter's presence even when we cannot directly observe it. Freeman says the lensing can reveal the properties of dark matter halos and lead to an estimate of the total mass of dark halos in the universe.—RA

"The Hunt for Dark Matter in Galaxies"
by Ken C. Freeman
Science, December 12, 2003

According to the current paradigm for galaxy formation, the early universe contained a mixture of baryons (ordinary matter made up from protons and neutrons) and cold dark matter (CDM) in a ratio of about 1:6 by mass. Small fluctuations in this mixture grew gravitationally into the galaxies, clusters of galaxies, superclusters, and filamentary large-scale structure of today's universe. The CDM model can reproduce most of the large-scale structure observed in the universe, but on galactic scales, there are some important discrepancies.[1] New advances in theory and observations are helping to resolve some of these discrepancies.

Galactic dark matter lies in dark halos that envelop the luminous parts of their parent galaxies. The motions of small satellite galaxies show that the halo of our own galaxy, the Milky Way, extends beyond about 300,000 light-years—much further than the galactic disk that contains most of the visible mass. The dark halo's mass is about 20 times that of all stars and gas in the galaxy.

According to the CDM model, large galaxies like the Milky Way should have large numbers of satellite galaxies; the Milky Way is expected to have about 500 small satellites. But only 20 or so are observed.[2] The model also predicts that the density distribution should rise rapidly toward the center of the halo, following a steeply cusped power law.[3] In contrast, most observations indicate that dark halos have a central core of nearly constant density.[4] Despite these differences, both theory and observation may be correct—for example, if the halos first form with many satellites and steeply cusped central regions, but star formation and expulsion of gas later help to unbind some of the satellites and flatten the central cores.

In most galaxies, the stars and gas rotate. The origin of their angular momentum, which is presumably shared by the dark matter, has long been thought to come from the tidal interactions of fluctuations in the early universe. Numerical simulations of this process do produce rotating systems, but the predicted internal distribution of angular momentum does not match the observations. Other sources of angular momentum may thus be needed. One such source may be the acquisition of angular momentum through accretion of smaller galaxies by larger ones.

Spiral galaxies do not usually rotate like rigid bodies. Their rotation curves (the variation of rotational velocity with radius) can be used to derive the typical densities and core sizes for dark halos, but first the gravitational field resulting from the dark halo alone must be estimated. At large distances from the galactic center, the gravitational fields are almost always dominated by the dark halo, but closer to the center, the stars (and gas) as well as the dark matter should contribute to the gravitational field. How can the gravitational contribution of the stars in these inner regions be calculated accurately?

Several methods have become available recently. One method uses stellar population synthesis models to estimate the surface density of the disk from its surface brightness. Another studies barred galaxies, in which the gravitational field of the bar (an elongated feature in the inner parts of the galaxy made primarily of stars) disturbs the rotation. The size of this disturbance, relative to the rotation, provides an independent estimate of the stellar gravitational field.

Although there is no complete agreement yet, most of these approaches give a consistent picture in which the halo dominates the gravitational field at large distances, but not in the inner regions of the galaxy. In barred galaxies, the angular momentum transfer between bar and halo can help the bar to grow; the disk becomes denser in this process, and as a result, it dominates the inner gravitational field.

Exciting new results about dark matter in galaxies are coming from another powerful method for measuring gravitational fields. As light travels toward us from a distant galaxy and passes by intervening galaxies, it is gravitationally deflected. This gravitational lensing effect can produce multiple images of the distant galaxy if the line of sight passes close to an intervening galaxy (strong lensing). The gravitational field of the dark halo of the intervening galaxy dominates the lensing. Sometimes, strong lensing results in anomalous intensity ratios among the multiple images of distant galaxies. The anomalies are probably generated by substructure in the dark halos of the intervening galaxies.

When the line of sight passes at larger distances from the intervening galaxy, the lensing generates only weak distortions in the observed shapes of the distant galaxies. Despite its weakness, this effect can be detected statistically from the huge samples of galaxy images that are becoming available from surveys. Weak lensing is an important tool for studying the properties of dark matter halos and for estimating the total mass of dark halos in the universe.

Combining the latest weak lensing data for dark halos with results from the WMAP cosmic microwave background mission[5] and the 2dF Galaxy Redshift Survey[6] yields a census of the different forms of mass in the universe. Stars and cold gas provide just 0.4 % of the mass of the universe. When gas in all forms is included, this percentage increases to 4 % for all baryons. Dark halos alone contribute about 11 %. All types of matter (dark matter plus baryons) add up to 27 %. The remaining 73 % comes in the form of a currently poorly understood "dark energy." The mass ratio of dark halos to stars and cold gas, 11 to 0.4, is close to the independent dynamical estimate of about 20 for the Milky Way.[7]

The largest galaxies have dark halos with masses of up to $\sim 10^{13}$ solar masses. At the other end of the scale are the dwarf spheroidal neighbors of the Milky Way and M31 (Andromeda). These very small and extremely faint galaxies have dark halos of $\sim 10^7$ solar masses. Their dark halos may represent the smallest quantum of galactic dark matter that could condense from the expanding universe.

Only a few percent of the mass of these galaxies is in the form of stars. In principle, galaxies with dark halos but no baryons could also exist, as could galaxies with dark halos and gas but no stars. However, deep neutral hydrogen surveys like the recent all-sky HI survey at Parkes[8] have found no examples of dark galaxies with gas but no stars. Current weak lensing surveys have not yet detected any indication of entirely dark galaxies.

Early in the evolution of the universe, theory predicts that the first objects to detach from the expanding universe were small and dense. We would expect to find correlations between the size and density of dark halos and the brightness of the galaxies that inhabit them. A careful reassessment of dark halo densities and sizes shows that these correlations are indeed present with the expected slope.[9] This is another indication that the basic current paradigm of galaxy formation is correct.

References

1. *Dark Matter in Galaxies*, International Astronomical Union Symposium 220, Sydney, Australia, 21 to 25 July 2003.
2. B. Moore *et al.*, *Astrophys J.* 524, L19 (1999).
3. J. Navarro, C. Frenk, S. White, *Astrophys J.* 462, 563 (1996).
4. W. de Blok, A. Bosma, *Astron. Astrophys.* 385, 816 (2002).
5. See http://antwrp.gsfc.nasa.gov/apod/ap030212.html.
6. See http://msowww.anu.edu.au/2dFGRS.
7. M. I.Wilkinson, N.W. Evans, *Mon. Not. R. Astron. Soc.* 310, 645 (1999).
8. See http://www.atnf.csiro.au/research/multibeam.
9. P. R. Shapiro, I.T. Iliev, *Astrophys. J.* 565, L1 (2002).

————   ————

In the next article, Robert Simcoe describes the tenuous structures of gas that reside in intergalactic space. Although exceedingly sparse, these outweigh all the known stars and galaxies by perhaps as much as 50 percent. Because of the enormous volume between the galaxies, the gas mass adds up quickly.

Recently, computer models simulating the appropriate physics show that the gas forms crisscrossing networks of "filamentary tendrils," a pattern nicknamed "cosmic webs." Before then, it was thought the gas resided in discrete, unconnected spheres.

Dark matter, Simcoe says, played a vital role in forming the webs. During the first 380,000 years of the universe, photon-related gas pressure in the primordial plasma kept ordinary matter from gravitationally condensing, but not dark matter, which isn't susceptible to the gravity-resisting pressure. When this period ended and the pressure abated, the dark matter structures provided ready gravitational anchors for the ordinary matter. *Galaxies also appear to form where several web tendrils intersect and provide feedback that alters the gas. Starlight heated up the colder intergalactic material and ionized it by stripping away electrons.*

Simcoe also discusses the "chemical pollution" of the webs caused when powerful particle winds bring heavy elements forged in supernova explosions. —RA

"The Cosmic Web"
by Robert A. Simcoe
American Scientist, January–February 2004

Observations and simulations of the intergalactic medium reveal the largest structures in the universe

There is no such thing as empty space. The idea of absolute emptiness realizes its closest approximation in the barren expanses between the stars and the galaxies, but even the most remote corners of the universe are suffused with very low density gas—which becomes increasingly rarefied as one ventures farther away from the places where galaxies consort. Consider this fact: In the air we breathe, each cubic centimeter contains roughly 5×10^{19} atoms. In contrast, the intergalactic medium has a density of only 10^{-6} particles per cubic centimeter—each atom inhabits a private box a meter on each side. This would seem to suggest that there is not much matter in the intergalactic medium. But, given the enormous volume between the galaxies, it quickly adds up: The combined atomic mass of intergalactic gas exceeds the combined atomic mass of all the stars and galaxies in the universe—possibly by as much as 50 percent! There is indeed something in empty space.

As cosmologists construct new narratives of the universe's evolution from its beginning—the Big Bang—to the present day, it is becoming clear that we must understand the physics of intergalactic matter if we are to write the history of how the galaxies, stars and planets formed. In the past decade, rapid advances in both the design of telescopes and computing power have allowed us to study the remote corners of intergalactic space in unprecedented detail. These new results deepen our understanding of how the grandest structures in the universe formed and evolved.

In the Red

Intergalactic gas is so tenuous and dark (producing no light of its own) that you might well ask how astronomers can hope to observe it. The trick is to detect it indirectly, by seeing how it influences light coming from faraway sources. The most common object for these observations is a quasar, a special type of galaxy containing a supermassive black hole at its center. Gas around the black hole emits intense radiation, which often outshines the average galaxy by 100 or more times. Because quasars are so bright, we can observe them at great distances and so measure the effects of intergalactic gas over substantial portions of the universe.

Using the world's most powerful telescopes, we can collect photons from these distant beacons and sort them by their wavelengths into spectra. The strongest feature in such a record is an emission line that is produced by hydrogen atoms near the quasar's black hole. The electrons in these atoms are excited to a single quantum level above their ground state. When they settle back to ground, photons are emitted with the precise wavelength of 121.56701 nanometers—called the Lyman-α transition. Yet we observe the emission line at a much longer wavelength, 560 nanometers. This is because the quasar is racing away from us, carried by the general expansion of the universe (see *American Scientist*, "The Hubble Constant and the Expanding Universe," January–February 2003). The expansion is such that objects far from us recede proportionally faster than those that are close. As an object moves away from us, the light that it emits is stretched to longer wavelengths in much the same way that the Doppler effect lowers the pitch of a receding train whistle. Astronomers use the term *redshift* to describe this phenomenon, since the colors of ever-more-distant objects become systematically redder.

Now consider what happens to the light of a quasar when it is transmitted through the intergalactic medium. As light from the quasar heads toward the Earth, some of its photons will intercept hydrogen atoms along the way. If one of these photons has a wavelength of 121.56701 nanometers, it will be absorbed by the atom, which then has one of its electrons kicked out of the ground state. When the electron loses energy and falls back to the ground state, the photon is re-emitted, but in an arbitrary direction, which is not likely to be toward Earth. So a cloud of hydrogen atoms will absorb light at a very specific wavelength and scatter it away—we see this as a dark "hole" in the spectrum.

The intergalactic medium contains many hydrogen clouds at different distances from us. And because clouds at different distances have different redshifts, a quasar spectrum shows many absorption lines at different wavelengths. The wavelengths below the hydrogen emission line thus appear to be "eaten" away according to the location of each cloud between the quasar and us. In the past decade new instruments on large telescopes have allowed us to examine the spectra of quasars at very-fine-wavelength resolution and high signal-to-noise ratio. These "zoomed-in" views resolve the intergalactic medium into individual clouds.

Spinning a Cosmic Web

When the absorption lines of quasars were first studied, it was not at all clear how to interpret them, particularly without the benefit of the high-quality data we have today. From the late 1970s to the early 1980s, Wallace Sargent's team at Palomar Observatory made a series of measurements that convinced most astronomers that these absorption lines represent intergalactic matter. However, a number of theoretical explanations were consistent with the available data, and most models explained the lines as clusters of discrete spherical clouds of gas.

In recent years, the advances in observing techniques have been joined by increasingly powerful computer models, which together deliver a more sophisticated picture of the intergalactic medium. This work involves several collaborations and requires months of supercomputer time. In these simulations, an imaginary box is designed to resemble a large representative volume of the universe. The box is divided up into a three-dimensional grid of cells, and matter is distributed throughout the grid in an initial state—according to conditions set by observations of the early universe. All of the physical processes that affect the evolution of the intergalactic medium are dialed into the model. Then the simulation is "turned on," allowing matter and energy to flow from cell to cell in the box, governed simply by the physics. The final product resembles a cosmic time-lapse movie with millions of years compressed into each frame. The computer code examines the distribution of matter in the box at each frame, or *time step*, and calculates the total force acting on each particle to determine where it should move in the next step. At regular intervals, the computer records the density of the gas throughout the intergalactic medium, and these results are compared with actual observations of quasar spectra to test the accuracy of the physical models.

One such output, from a simulation run by Jeremiah Ostriker and Renyue Cen of Princeton University, is shown in the top panel of Figure 1 [in the original article]. This particular view shows the universe when it was about 15 percent of its present age, or about 2 billion years old. The most striking feature seen is a tendency for gas to collapse into a network of filamentary tendrils that crisscross through vast, low-density voids. This pattern is a common feature of the new computational models and has been nicknamed "the cosmic web."

To test this depiction of the universe against concrete observations, large numbers of artificial quasar spectra are generated by drawing random lines through the simulation box. By evaluating the variations of gas density along any single line, astronomers can calculate the amount of absorption that would be observed in a spectrum measured along that line of sight. It is as though an observer stood on one side of the box and measured the spectrum of a quasar on the other side.

Statistically, the "spectra" from these artificial universes are nearly indistinguishable from the spectra of real quasars. The models accurately predict the number of absorption lines, the distribution of their strengths and widths, and their evolution through time. At a basic level, these models have captured the physical processes that dominate the evolution of the universe on the largest scales.

Lumps and All

Technology has thus given us the tools to observe the remote corners of the intergalactic medium and to interpret these observations in the context of a cosmological model. Having described the methods, now let us step back to examine the model itself by offering a narrative that explains the formation of galaxies and intergalactic structure.

The story begins more than 13 billion years ago, roughly 380,000 years after the Big Bang, when the universe was very different from today. There were no stars, galaxies or webs yet, just a uniform soup of free-floating protons and electrons. In fact, the gas was so evenly distributed that its peak densities differed by only 1 part in 100,000 from the cosmic average. But sometime between then and now it evolved into a very lumpy place, where vast stretches of nearly empty space are interrupted by "dense" strands of galaxies and gas. Today, the range of densities is much greater: The difference between the atomic density of the Sun's interior and intergalactic space spans about 32 orders of magnitude!

Astronomers believe that this transition from smooth to lumpy was driven by gravity. Imagine a box containing a perfectly uniform distribution of matter, so that the density of the particles is constant. Suppose that at one location in the box the particles are somehow stirred, leading to a slight density enhancement at this particular spot. This tiny new concentration of mass will create a gravitational force, which tugs on the surrounding particles and causes them to fall inward. The infalling matter increases the clump's mass, which in turn increases its gravitational pull, allowing it to assemble even more material, and so on. Given enough time, this "gravitational runaway" transforms what was originally a tiny density enhancement into a dense clump, containing most of the mass that was distributed throughout the volume.

This simple phenomenon is the basis for theories of how the large-scale structure of the universe was formed. Yet in order for it to work, the universe must have been "imprinted" at some earlier time with a network of primordial density perturbations that would later collapse into the structures we see today. As it happens, the signature of these ripples has been observed—as tiny variations in the temperature distribution of microwave photons coming from different parts of the sky. Characterization of this microwave background is currently a major focus of astronomical research, as the ripples represent the ancient gravitational seeds of cosmic structure.

It would seem that we have all the elements needed to explain the origin of the cosmic web. We have observed density variations in the early universe, and we have a powerful model that explains how they could evolve into larger structures. However, there is one problem: The primordial variations were so small that 13.7 billion years is still not enough time to grow them into the assemblages we observe today! This puzzle received a great deal of attention during the 1970s, perhaps fueled by Cold War politics. Two competing theories of structure formation emerged, one devised by Yakov Zel'dovich at the School of Russian Astrophysics in Moscow, and the other by James Peebles and his collaborators at Princeton University. The ensuing debate exposed significant weaknesses in both theories. The solution required the introduction of an entirely new ingredient—ominously named *dark matter*—in the cosmological models. This proved to be one of the most important discoveries in modern cosmology.

This dark stuff is quite different from the ordinary matter that makes up stars, planets and people. Not only does dark matter not shine, it interacts with "our" kind of matter only through the force of gravity. It is largely believed to consist of exotic particles that have no other effects on ordinary atoms and molecules. Furthermore, dark matter appears to outweigh normal matter throughout the universe by a factor of four to one. This notion is indeed odd, and it has met with resistance since it was first suggested by the eccentric astronomer Fritz Zwicky in the 1930s. However, cosmologists have now grown to accept its existence as nearly certain in the face of overwhelming evidence from a variety of observations. Although we may not understand exactly what dark matter is, we do understand what it *does*—it holds galaxies together, bends light, slows down the universe's expansion and drives the formation of inter-galactic structure.

To understand this last point, we need to return to the early history of the universe. During the first 380,000 years, the relic heat from the Big Bang kept the universe so hot (greater than 3,000 kelvins) that electrons and protons in the primordial soup could not combine to form neutral hydrogen atoms. Such ionized gas, in this case consisting of dissociated electrons and protons, is known as a *plasma*. When plasma particles are in their free-floating state, they can interact with light, exchanging energy and momentum. In the early universe, this scattering increased the gas pressure within the cosmic soup. So, when gravity tried to collapse the first density perturbations, the gas pressure pushed back—much as a balloon does when it is squeezed. As long as the electrons and protons were separated, the gas could not form larger structures. Instead, the potential structures churned and oscillated as the inward pull of gravity fought the outward push of gas pressure.

Then, when the universe was 380,000 years old, a major event took place. As the universe was expanding, it was also cooling, and at this point it became cold enough for electrons and protons to combine, forming hydrogen atoms. Suddenly, these new atoms became decoupled from the photons—they no longer interacted so strongly with light—which drastically reduced the pressure that had kept gravity at bay. With gravity free to work on all the newly formed hydrogen atoms, structures could form in earnest.

How did dark matter fit into the picture? While the protons, electrons and photons were oscillating under the competing influences of gravity and pressure, the dark matter followed a different storyline. Because dark matter interacts with normal matter only through gravity, the pressure that kept the normal gas from collapsing couldn't act on it. Particles of dark matter enjoyed an unimpeded assembly into large structures long before the normal gas could begin to get organized. By the time normal matter decoupled from the photons, the dark matter had already grown into a primitive web-like network. As soon as the normal matter lost its support from the photon pressure, the gravity from the pre-existing dark-matter structures quickly pulled normal gas into the web. In this way, normal matter was given a gravitational "head-start" by the dark matter.

Once this process was set in motion, the gravitational building blocks of the intergalactic medium were in place. Normal and dark matter continued to free-fall toward concentrations of mass until the rising gas pressure slowed the infall. The web-like lattice was taking shape, but stars had not yet begun to form and all of the gas in the universe was neutral. The universe had entered an age where matter drifted about in the darkness, quietly assembling under gravity's influence. So it continued until at some point—probably somewhere between 200 million years and one billion years after the Big Bang—a process began that would fundamentally alter the nature of the intergalactic medium and the universe as a whole: The first stars were born.

Fiat Lux

It seems preposterous that something as small as a star could affect the universe on intergalactic scales. After all, a star is only a few light seconds across, whereas the filaments of the cosmic web may extend for billions of light-years. How can a relatively tiny object impact such a large volume? The answer lies in how stars work, where they live and what happens when they die.

Before there were stars, the normal matter in the universe was composed almost entirely of hydrogen and helium. Astronomers refer to this mixture as a chemically pristine gas because it reflects the chemical composition of the cosmos just after the Big Bang. Since then, nearly every atom of every other element—from argon to zinc—was forged inside a star. Stars are effectively nuclear fusion reactors: They gravitationally compress gas to such high densities that light atomic nuclei smash together to form heavier elements. Such stellar nucleosynthesis releases enormous amounts of energy, and that's what makes the stars shine.

Nucleosynthesis had several important effects on the intergalactic medium. First, it generated starlight, which escaped into intergalactic space and interacted with the neutral atoms. Later, the newly minted heavy elements were ejected into the intergalactic medium by strong *galactic winds*—powerful expulsions of hot gas—that stirred up and "polluted" vast regions of the universe.

Let's consider these processes in more detail by returning to the cosmic web. Because galaxies are more than 10,000 times denser than the cosmic average, we would expect to find systems like the Milky Way within dense regions of the web itself, which contain the raw materials (gas reservoirs) needed to build the stars and galaxies.

In simulations, the densest regions are found within the web's filaments, especially where several intersect. Therefore, on cosmic scales, galaxies should behave like tiny particles trapped in the strands of the web, actually tracing the much larger structures outlined by intergalactic gas. Recent three-dimensional galaxy surveys, such as the Sloan Digital Sky Survey and the 2dF Galaxy Redshift Survey, have indeed revealed a filamentary pattern in the way that galaxies cluster. Research groups, led by Max Tegmark at the University of Pennsylvania and Rupert Croft at Carnegie Mellon University, are currently investigating the clustering statistics of galaxies relative to those of the intergalactic gas as seen in quasar spectra. Their early results suggest that the same physics underlies the assembly of the intergalactic gas network and large-scale galaxy structures.

As the galaxies coalesced out of the web and began to shine, the universe was filled with the first new light since the Big Bang—the dark era had ended. And the stars dutifully began to churn out heavy elements. When enough stars had formed, the cumulative production of light and chemicals began to alter the nature of the intergalactic medium itself. Astronomers refer to these collective effects as "galaxy feedback," because the galaxies act on the surroundings from which they formed. Here I'll only consider two types of feedback, radiation and chemical pollution.

The first agent of galaxy feedback was starlight, which reionized the intergalactic medium. Recall that normal matter began to form large structures during the era of *recombination*, when the protons and electrons teamed up to form hydrogen atoms, so the gas in the universe was, for a time, entirely neutral. It was also very cold, reaching gas temperatures only a few tens of degrees above absolute zero. When the first stellar photons leaked out from galaxies, they interacted with the hydrogen atoms, stripping away the electrons that had been in place since the era of recombination and reheating the resulting plasma up to temperatures near 10,000 kelvins. Reionization was initially confined within bubbles centered on the fledgling galaxies, because the starlight had not yet traveled far out into intergalactic space. As more galaxies began to shine, the ionized bubbles grew outward until those from adjacent galaxies began to overlap. Soon the entire volume of the universe was once again ionized.

We now believe that the universe finally emerged from its "dark ages" and was reionized when it was less than 1 billion years old, or about 10 percent of its present age. Today, only about 1 hydrogen atom in 10,000 is in a neutral state and the average temperature of intergalactic gas is still very near 10,000 kelvins.

A Mighty Wind

It had long been assumed that the intergalactic medium was chemically pristine and that the production and distribution of new elements took place only within galaxies themselves. But astronomers also noticed that a few weak absorption features in quasar spectra appear redward of the hydrogen emission line. These other lines arise from different elements—in the case of Figure 2 [in the original article], carbon and silicon—whose characteristic wavelengths are redder (longer) than hydrogen's 121.56701 nanometers.

The absorption lines of these heavy elements are observed within regions that also contain a considerable amount of hydrogen. These zones correspond to gaseous halos around the first galaxies, whose stars were thought to supply the chemicals. However, in the early 1990s, quasar spectra taken by Lennox Cowie and Antoinette Songaila on the newly commissioned Keck telescopes revealed heavy elements far removed from any galaxy. Their discovery suggested that the chemical pollution of intergalactic space was much more efficient than originally believed.

The concentration of heavy elements in the intergalactic medium is very low: For example, only about one carbon atom can be found for every million (mostly hydrogen) atoms. So a box of intergalactic space that is 100 meters on a side would contain just a single carbon atom! Yet even this tiny amount reveals that some heavy elements were mixed throughout the cosmic web early in the history of the universe. How did they get out there—so far from the stars and galaxies in which they were made?

The evidence suggests that they were blown out into intergalactic space by violent *galactic winds*. These streams of matter flow out of galaxies where stars are actively forming. In all galaxies, the most massive stars burn brightly, and rapidly produce new elements. These stars burn so fast that they quickly exhaust their nuclear fuel and can no longer continue fusing light elements into heavier ones. When the reactor in a massive star turns off, the star ends its life in a tremendous explosion known as a supernova. The blast energy of a typical supernova rivals the simultaneous detonation of 10^{31} atom bombs, and the remnants of the dead star—including its newly fused heavy elements—are launched into surrounding space.

Despite its explosive power, a single supernova cannot pollute the intergalactic medium because the gravitational force from the star's galaxy traps the expanding debris before it can escape. However, galaxies occasionally experience bursts of unusually vigorous star formation where stars are born and die 10 to 50 times faster than usual. During these *starbursts*, multiple supernovae can be triggered in near succession. Their collective energy drives debris outward, like a rocket boosted by several stages, breaching the gravitational barrier and expelling heavy elements into the intergalactic medium. This phenomenon has been observed in a number of nearby galaxies.

Although we can study nearby starbursts and the resulting outflows in exquisite detail, these galaxies are the rare exception in the local universe. Most galaxies quietly go about forming stars and manage to retain the heavy elements they produce. But in the early universe, the situation was quite different. New observations of distant galaxies by Max Pettini at the University of Cambridge and his colleagues have revealed that outflows were extremely common when the universe was about 15 percent of its present age. This has two important implications. Nearly every galaxy we see today underwent some period of intense star formation in its past. And large quantities of heavy elements were launched into the intergalactic medium very early in the life of the universe. There was thus plenty of time for this material to coast out to large distances and mix with the chemically pristine intergalactic gas.

Studies of early galaxies and their feedback on the intergalactic medium define an important frontier of our knowledge about the first stars and cosmic structures. Several important questions remain open. For example, exactly when and where did the first stars form? Do heavy elements pervade the entire universe, or is there still chemically pristine gas left over from the Big Bang? Were the stars that triggered reionization the same stars that produced the observed intergalactic heavy elements?

For the past few years I have been investigating some of these questions with Wallace Sargent at the California Institute of Technology and Michael Rauch at the Carnegie Observatories. We have been measuring heavy-element concentrations in the early cosmic web to learn whether there are pristine corners of the universe that have not yet been reached by the galactic winds. So far we have detected heavy elements throughout all of the strands of the cosmic web, but it is still not clear whether the winds' sphere of influence extends beyond the filaments and into the intergalactic voids. In these remote regions the expected heavy-element densities are so low that even our most sensitive observations cannot reveal their absorption lines directly. Nevertheless, our results show that debris from galactic winds must have dispersed into most of the mass in the universe before the cosmos was a mere 20 percent of its present age.

We have also compared our observations with different models of star formation and chemical production to determine whether the stars that reionized the universe were the same ones that polluted the intergalactic medium. Our results suggest that the earliest stars did not produce most of the heavy elements, most likely because their heyday was too short. Instead, we believe that galaxy feedback occurred in a series of waves. The first generation of stars reionized the universe, and later generations progressively enriched the intergalactic medium with chemicals.

On the theoretical front, the most advanced cosmological simulations are just beginning to incorporate
realistic models of galactic winds and the chemical
enrichment of the universe. The physics of star formation
and galaxy outflows is so complex that even the most
sophisticated numerical models must make broad, simpli-
fying assumptions to make the problem computationally
tractable. The subject continues to progress rapidly, as
both the observations and the theory evolve.

There are, of course, many details to be refined.
Exactly how and when did the first stars form? How do
galaxies and the intergalactic medium interact? And,
perhaps most importantly, what is the nature of dark
matter? Yet, when sufficient time has passed to offer a
historical perspective, the past decade may well be
remembered for the emergence of a standard model of
the cosmos that ties all we know about galaxies and the
intergalactic medium into a single package.

Reprinted with permission from *American Scientist.*

Although we have a decent grasp of how chemical elements are created through nucleo-synthesis, their origin and history are far from clear, writes author Max Pettini in the next article. But the unexpected presence of heavy elements detected in a far distant galaxy is helping astronomers unravel these unknowns. The heavy elements imply a surprising amount of star formation in a galaxy far from our Milky Way galaxy, when the universe was just 1 or 2 billion years old.

Elements let themselves be known by their electromagnetic emissions, caused when electrons drop to lower energy states, and also by the radiation they absorb, which boosts electrons to higher energy states. However, because the feeble light of this far-off galaxy could not be detected, researchers relied instead on the absorption of intense quasar light shining through it, which required the fortuitous alignment of the quasar and the galaxy. The detection of elements as heavy as tin and lead tells Pettini that "galaxies formed stars at different rates and at different times," so the underlying physics then and now must be essentially the same.

"The importance of such a link between the past and the present . . . is hard to overestimate," he says. —RA

"Distant Elements of Surprise"
by Max Pettini
Physics World, July 2003

The fortuitous alignment of a quasar and a distant galaxy has enabled astronomers to unravel the origin and evolution of chemical elements

Cosmologists and fossil hunters have more in common than it might first appear. Palaeontologists analyse fossil records of the last 3.5 billion years to track the evolution of life on Earth. Motivated by a similar curiosity, astronomers are searching for chemical elements in stars and cosmic gas to understand the origin of the chemistry of the universe, and how it has developed since the Big Bang.

The basic theoretical framework of the universe's chemistry has been in place for half a century. The lightest elements—hydrogen and helium, along with trace amounts of deuterium and lithium—were created within a few minutes of the Big Bang. All of the other elements were assembled at much later times by the fusion of hydrogen and helium nuclei in the interior of stars. Some of these elements were dispersed through space in the violent supernova explosions with which some stars end their lives. They were then condensed into subsequent generations of stars, and into the planets that formed around them. Some of these elements were eventually incorporated into organic structures and became essential for life on Earth.

This cosmic cycle continues to the present day. The closest region of active star formation is the Orion nebula, which is visible to anyone who looks up at the sky on a clear winter's night. But the details of the cycle are still far from clear. For example, we do not know when most of the elements were produced in the universe as a whole, or which stars were responsible for producing which elements. In order to unravel the path of chemical evolution we have to shift our attention to much further away than the Orion nebula.

Lucky Alignment

Traditionally, astronomers have looked for clues about stellar chemistry in stars of different ages and at different locations within our galaxy—the Milky Way. Stars in the halo of the Milky Way probably formed a very long time ago. Those that are still around us bear the imprint of the chemical composition of the proto-galaxy in their atmospheres, like living fossils of a bygone era in the history of the Earth. But this is only part of the story. The Milky Way is just one of several billion galaxies and its stellar population therefore gives us only a limited window on the process of galactic chemical evolution.

Thanks to recent technological advances, astronomers have now begun to extend their studies of chemical abundances beyond the Milky Way. In a major development, we are now able to see nearly all the way back to the Big Bang. Large telescopes, such as the Very Large Telescope at the European Southern Observatory in Chile and the Keck telescopes on Hawaii, are so efficient at gathering and recording light that astronomers are able to study galaxies at the edge of the observable universe.

The light that falls on these telescopes left some galaxies when the universe was just 1 or 2 billion years old. We know this from its redshift—an increase in the apparent wavelength of the light due to the expansion of the universe. These galaxies therefore appear not as they are today, but as they were 12 billion years ago— just what the galactic fossil hunters are looking for.

Some of the galaxies can be seen directly, while others are only observable by the shadow that they cast in the light of more distant and brighter objects. These two techniques for studying high-redshift galaxies are highly complementary.

When a galaxy is bright enough that its starlight can be seen directly, we can use spectroscopy to discern its chemical composition and are able to relate it to nearby galaxies around us today. However, the furthest galaxies are generally too faint to reveal their chemical composition in detail. For this we need to use high-resolution spectroscopy. But this is only possible when the light from something much brighter—such as a quasar—shines through the galaxy, thanks to a chance alignment of the two objects as seen from Earth.

One such alignment was recently discovered by a team of astronomers in the US. Jason Prochaska of the University of California at Santa Cruz, and Chris Howk and Artie Wolfe at the University of California in San Diego have been systematically searching for such galaxy-quasar pairs and following up the most promising candidates with the Keck telescopes. This is the first pair that they have found in over a 100 candidates (see *Nature*, vol. 423, pages 57-59).

The pair consists of a quasar that shines through a dense region in a foreground galaxy, which means that the number of atoms available to absorb the quasar light is unusually large. The spectral signatures of the quasar occur in a different part of the spectrum to those of the foreground galaxy due to their different redshifts. The spectral lines of the quasar are also broader than those of the intervening galaxy because they are produced by a hot, as opposed to the cold galactic, gas. The team found that the intervening galaxy is surprisingly advanced along the path of chemical evolution.

Elemental Abundance

The redshift of the galaxy is 2.626—which corresponds to looking back in time some 12 billion years. Stars in our galaxy that are this old contain only small amounts of chemical elements, typically just one-tenth to one-hundredth of the amounts in the Sun. This is generally taken as a sign that little star formation took place in the proto-Milky Way.

But we now have a quite different example of a galaxy that underwent a significant amount of star formation in just 2.5 billion years, during which its chemical composition became similar to that of the Sun. In particular, the abundance of oxygen in the stars of this galaxy had already grown to one-third of the solar value.

The Milky Way is a spiral galaxy and observations show that in most "spirals," star formation has proceeded at a relatively steady pace over their entire existence. But there are other galaxies in the nearby universe that astronomers suspect have squandered most of their resources in their youth, by forming stars at much faster rates. These are the elliptical galaxies, which have very little gas left and contain a predominantly old collection of stars.

Prochaska and colleagues may well have found the progenitor of one of today's elliptical galaxies, and observed it at an early stage of its evolution. The importance of such a link between the past and the present, and between different classes of galaxies, is hard to overestimate.

It tells us that galaxies formed stars at different rates and at different times. Some formed slowly and at an even pace, while others were born in one major episode of star formation that took place a long time ago. This is what we had suspected from the way galaxies look today, but Prochaska and co-workers have found a concrete example that confirms some of these ideas.

Furthermore, the rare combination of high gas density and high element abundances in this particular quasar-galaxy pair produces an extraordinarily rich spectrum. This brings within reach a number of atomic transitions that are normally far too weak to be detected. Prochaska's team has measured 25 chemical elements in this distant galaxy ranging from the relatively light elements boron and nitrogen to real heavyweights such as tin and lead.

Extragalactic Heat

Many of these elements had never been seen before outside the Milky Way. They therefore provide an unprecedented opportunity to study the complex pattern of element abundances in a galaxy that is in a very different place and time in the universe.

This pattern is surprisingly similar to that in the Sun, which is a strong clue that the same basic physical processes that synthesize chemical elements inside stars here and now were operating there and then. It also means that some universal law—which is roughly invariant in time and space—must determine the relative numbers of stars of different masses that form in a galaxy. Stars that have different masses produce different amounts of each chemical element, yet the final mix seems to be approximately the same. Since stars of different masses synthesize different elements in different proportions, this need not be the case.

The challenge now is to extend this type of measurement to galaxies with even higher redshifts. This will allow astronomers to look for chemical traces that were left over from the very first generation of stars to form in the universe—perhaps only a few 100 million years after the Big Bang.

These redshifts are so high that the wavelength of the starlight will be shifted not into the visible region—as in the galaxy studied by Prochaska and collaborators—but all the way into the infrared. Advanced plans are already in place to construct near-infrared spectrographs for most large telescopes, and the first of these will be in operation in just a couple of years.

Reprinted with permission from *Physics World*.

There's an old cliché that says the universe can be seen in a grain of sand. In the next article, the authors make a more modest claim—that they can see the birth of the solar system in the grains of a meteorite. They propose a model that could rewrite our understanding of the early history of our small corner of the universe.

Conventional wisdom "is almost certainly wrong," they write, when it says our solar system came from a sedate setting like "the dark interior of an isolated molecular cloud." Rather, it came about in the "far more violent environment around the periphery" of a region of hydrogen gas ionized by the ultraviolet radiation from nearby hot, massive stars, and where at least one star went supernova.

The key to this model is the presence of a special nickel isotope in a type of meteorite that undergoes little alteration after formation in the early solar system. The nickel isotope could only have come from the radioactive decay of a short-lived iron isotope, [60]Fe, in abundances that recent work shows could have only been created in a supernova explosion. —RA

"The Cradle of the Solar System"
by J. Jeff Hester, Steven J. Desch, Kevin R. Healy, and Laurie A. Leshin
Science, May 21, 2004

What kind of environment gave birth to the Sun and planets? Most astronomers who study star formation would probably say that the solar system originated in a region much like the well-studied Taurus-Auriga molecular cloud[1]—a region in which low-mass, Sun-like stars form in relative isolation—but this conventional wisdom is almost certainly incorrect. Recent studies of meteorites confirm the presence of live [60]Fe in the early solar system.[2] No known mechanism could have formed this short-lived (half-life = 1.5 million years) radionuclide locally within the young solar system. However, [60]Fe is produced in supernova explosions, along with [26]Al, [41]Ca, and other radioisotopes.[3] Material from nearby supernovae must have rapidly mixed with the material from which the meteorites formed. The implications of this are clear. The Sun did not form in a region like Taurus-Auriga. Rather, like most low-mass stars,[4] the Sun formed in a high-mass star-forming region where one or more stars went supernova. Understanding our origins *means* understanding the process of low-mass star formation in environments that are shaped by the presence of massive stars.

The intense ultraviolet (UV) radiation from massive stars carves out ionized cavities and blisters in the dense molecular clouds within which the stars formed. Examples of these regions of ionized gas, called HII regions, include such well-known objects as the Orion Nebula and the Eagle Nebula. There is growing evidence that most low-mass star formation in such environments is triggered by shocks driven in advance of the HII region ionization front as it expands into its dense surroundings.[5] Stars seen in the ionized volumes of HII regions were formed in this way, and then subsequently were uncovered by the advance of the ionization front itself.

Low-mass stars that form around an HII region should pass through a well-defined sequence: (i) A shock driven in advance of an ionization front compresses molecular gas around the periphery of an HII region, compressing dense cores and causing them to become unstable to gravitational collapse.[6] (ii) These cores are overrun by the advancing ionization front within $\sim 10^5$ years. As cores emerge into the HII region interior, they go through a short-lived ($\sim 10^4$ years) phase during which the dense core itself photoevaporates. This is the "evaporating gaseous globule" or EGG phase best seen in Hubble Space Telescope (HST) images of the Eagle Nebula.[7] (iii) EGGs that do not contain stars are dispersed, but when a star-bearing EGG evaporates, the circumstellar disk inside is exposed directly to UV radiation from the massive stars. The object transitions into an "evaporating disk" phase, best seen in HST images of "proplyds" in the Orion Nebula.[8] (iv) The evaporating disk phase is also short-lived.[9] Within a few tens of thousands of years, photoevaporation erodes the gaseous disk to within a few tens of astronomical units of the central young stellar object (YSO).[10] (v) The young star and its truncated disk then reside within the ionized, low-density interior of the HII region for the remainder of the few-million-year lifetime of the region. This is the environment in which planetary systems such as our own form. (vi) When the massive stars exciting the region go through a high mass-loss "Wolf-Rayet" phase and/or go supernova, the protoplanetary disks surrounding nearby low-mass YSOs are pelted with ejecta. Such events are responsible for the short-lived radionuclides found in meteorites in our own solar system.

This scenario for star formation makes many testable predictions that are supported by observations already in the literature. For example, fingerprints of the star-formation process discussed here are clear in the HST image of a region in the Trifid Nebula[11] shown in the figure [in the original article]. In this region, intense UV radiation from a massive star (located well above the field of view) is incident on the surface of dense molecular gas that fills most of the field of view. Sharply defined orange and yellow features mark the current location of the ionization front. The HH399 jet originates from an unseen protostar located a short distance from the ionization front. A water maser is also seen in projection a short distance behind an ionization front. Jet and maser activity are both evidence of continuing accretion onto these two very young protostars. In 10,000 years or so, both of these objects will be cut off from their accretion reservoirs when they are overrun by the advancing ionization front. When this happens, these objects will be seen as EGGs, much like the prominent EGG shown in the inset [in the original article].

The EGG seen in the figure is itself a remarkable demonstration of the evolutionary tie between EGGs and proplyds. From the bottom down, this feature is a classical EGG, of the sort seen in the Eagle Nebula. But at the tip of this EGG we see a star, a small reflection nebula, a small protostellar jet, and an ionization front in the evaporative flow off the tip of the structure. These features are all characteristic of the proplyds seen in the Orion Nebula.[8] In other words, this object is undergoing the transition from EGG to proplyd.

One of the clearest predictions of this scenario is that star formation is a sequential process. There should be a clear relationship between the properties of YSOs and their distance from the ionization front. Many lines of evidence confirm this prediction. For example, Hα-bright protostars are known to be concentrated near ionization fronts in numerous HII regions,[12] as are water masers and other tracers of star formation.[5] The sequential nature of star formation is also apparent in the figure, where the protostars within the ionized volume of the HII region are clustered into several small groups that were left behind by the advance of the ionization front. One such group in the upper left portion of the field is especially telling, because these stars still surround a small molecular teardrop—the remnant of the larger molecular core that gave birth to these stars and was subsequently evaporated by the advance of the ionization front. This is what the adjacent TC2 molecular column will look like in $\sim$ 100,000 years.

Most low-mass stars and planetary systems, including our own, formed in HII region environments. A unified description of this orderly process has broad implications. For example, the initial distribution of stellar mass is largely a consequence of this process. Closer to home, the solar system formed from a truncated disk bathed in intense UV radiation from massive stars and subjected to the effects of nearby supernovae. This scenario has consequences for questions as diverse as the truncated outer edge of the Kuiper Belt, oxygen-isotope anomalies in meteorites, and the differentiation of planetesimals driven by decay of short-lived radionuclides. The fields of astrophysics, meteoritics, astrobiology, and planetary science meet in the early solar system. The setting for that encounter is not the dark interior of an isolated molecular cloud, but rather the far more violent environment around the periphery of an HII region. The predictive scenario for the origin of low-mass stars proposed here, with its roots in the study of both meteorites and star formation, provides a context and direction for future work in each of these fields, and in the theory that unites them.

References

1. F. C. Adams, G. Laughlin, *Icarus* 150, 151 (2001).
2. S. Tachibana, G. R. Huss, *Astrophys. J.* 588, L41 (2003).
3. B. S. Meyer, D. D. Clayton, *Space Sci. Rev.* 92, 133 (2000).
4. C. L. Lada, E. A. Lada, *Annu. Rev. Astron. Astrophys.* 41, 57 (2003).
5. K. R. Healy, J. J. Hester, M. J. Claussen, *Astrophys. J.*, in press.
6. F. Bertoldi, C. F. McKee, *Astrophys. J.* 354, 529 (1990).
7. J. J. Hester et al., *Astron. J.* 111, 2349 (1996).

8.   J. Bally, C. R. O'Dell, M. J. McCaughrean, *Astron. J.* 119, 2919 (2000).
9.   J. Bally et al., *Astron. J.* 116, 854 (1998).
10. D. Johnstone, D. Hollenbach, J. Bally, *Astrophys. J.* 499, 758 (1998).
11. J. J. Hester *et al.*, *Bull. Am. Astron. Soc.* 31, 932 (1999).
12. K. Sugitani *et al.*, *Astrophys. J.* 565, L25 (2002).

The universe contains billions of galaxies, each containing billions of stars. The surmise that some of these stars must have planets is being confirmed as better instruments and new techniques sleuth them out. But despite confirming the existence of more than 100 planets, we still don't quite know how they formed.

This shortcoming may be addressed by Beta Pictoris, a young star sixty-three light-years away, considered by some to be the "missing link" between young stellar objects with preplanetary disks and more highly evolved systems with formed planets. In 1984, astronomers discovered a dusty ring orbiting the star at 30 astronomical units (AU) (Earth orbits the Sun at 1 AU), and in 2003 found another at 16 AU.

A year later, as Ron Cowen reports in the following article, a high-resolution survey confirmed the second ring and discovered a third composed of fine silicate particles at 6.4 AU. In addition, the innermost rings likely have a planet between them.

Also, the stony material that makes up the rings is constantly being replenished, probably through the collision of bigger bodies, because the pressure exerted by the sunlight would drive the small particles away. —RA

"Planet Signs? Sifting a Dusty Disk"
by Ron Cowen
Science News, October 9, 2004

Twenty years ago, astronomers peering at the young star Beta Pictoris got their first glimpse of a disk of dusty debris—the sign that planets, asteroids, and comets are forming and then banging together and releasing an abundance of dust.

Such debris disks have now been found around many young stars, but the one surrounding Beta Pictoris remains the most revealing about how planets form and evolve.

Recording spectra of the disk at midinfrared wavelengths, Yoshiko K. Okamoto of Ibaraki University in Japan and his colleagues now have gleaned new details about the size, composition, and crystal structure of dust particles. The data, which indicate three distinct bands of dust within the Beta Pictoris disk, suggest the location of a possible planet as well as of a trio of asteroid or comet belts. The astronomers describe their study in the Oct. 7 *Nature*.

The team used an infrared camera on the 8.2-meter Subaru Telescope on Hawaii's Mauna Kea. Okamoto and his collaborators found that small dust particles without a crystal structure—such as amorphous silicate grains—collect into belts within the Beta Pictoris disk. These dust bands reside at 6.4, 16, and 30 astronomical units (AU) from the star. An AU is the distance between the sun and Earth, or roughly 150 million kilometers. Images taken by other researchers had discerned the outermost two bands, but the new study is the first to reveal their composition, Okamoto notes.

It takes less than 100 years for the pressure exerted by photons streaming from Beta Pictoris to blow tiny dust particles from the disk into more rarefied regions of space. The persistence of the particles in the belts suggests that they are continually replenished, probably by the evaporation of comets or the collision of asteroids, says Okamoto. The team's observations, which reveal that very few fine-grain silicates lie close to the star, favor the asteroid scenario, he adds.

After analyzing the forces exerted on the three dust bands, the team concluded that the tug of an unseen planet may have kept the belts intact over millions of years. The proposed planet would lie at 12 AU from the star, slightly beyond Saturn's distance from the sun.

The team also reports that the inner part of the Beta Pictoris disk, within a few AU of the star, has a different composition from material that lies farther out. The crystal form of the mineral olivine is more concentrated near the star, as are silicate grains with diameters greater than a few micrometers.

Alycia J. Weinberger of the Carnegie Institution of Washington (D.C.) says she's impressed by the mapping of the minerals in the star's disk. For grains to acquire a crystal structure, they must be heated to several thousand kelvins, she notes. That typically occurs extremely close to a star such as Beta Pictoris—no farther than about 0.1 AU, or one-fourth Mercury's separation from the sun. Finding crystals at greater distances than that, as Okamoto's team did, suggests that some force is mixing the material in the disk.

Mapping the composition of debris disks may eventually explain "what determines the composition of the planets," says Weinberger. The Beta Pictoris observations provide a "new piece of the puzzle of how solar systems and Earthlike planets form," says Steve Desch of Arizona State University in Tempe.

Web Sites

Due to the changing nature of Internet links, the Rosen Publishing Group, Inc., has developed an online list of Web sites related to the subject of this book. This site is updated regularly. Please use this link to access the list:

http://www.rosenlinks.com/cdfa/sistu

For Further Reading

Ferris, Timothy. *The Whole Shebang: A State-of-the-Universe(s) Report*. New York, NY: Simon & Schuster, 1997.

Fox, Karen C. *The Big Bang Theory: What It Is, Where It Came From, and Why It Works*. New York, NY: Wiley, 2002.

Greene, Brian. *The Fabric of the Cosmos: Space, Time, and the Texture of Reality.* New York, NY: A. A. Knopf, 2004.

Guth, Alan H. *The Inflationary Universe: The Quest for a New Theory of Cosmic Origins.* Reading, MA: Addison-Wesley Publishing, 1997.

Kirshner, Robert P. *The Extravagant Universe: Exploding Stars, Dark Energy, and the Accelerating Cosmos.* Princeton, NJ: Princeton University Press, 2002.

Lemonick, Michael D. *Echo of the Big Bang.* Princeton, NJ: Princeton University Press, 2003.

Levin, Janna. *How the Universe Got Its Spots: Diary of a Finite Time in a Finite Space.* Princeton, NJ: Princeton University Press, 2002.

Seife, Charles. *Alpha and Omega: The Search for the Beginning and End of the Universe.* New York, NY: Viking, 2003.

Villard, Ray, and Lynette R. Cook. *Infinite Worlds: An Illustrated Voyage to Planets Beyond Our Sun.* Berkeley, CA: University of California Press, 2005.

Wilford, John Noble, ed. *Cosmic Dispatches: The New York Times Reports on Astronomy and Cosmology.* Rev. Ed. New York, NY: W. W. Norton, 2002.

Bibliography

Board on Physics and Astronomy. "Executive Summary." *Connecting Quarks with the Cosmos: Eleven Science Questions for the New Century.* Washington, D.C.: The National Academies Press, 2003.

Cho, Adrian. "Galaxy Clusters Bear Witness to Universal Speed-Up." *Science*, Vol. 304, No. 5674, May 21, 2004, p. 1,092.

Cline, David B. "The Search for Dark Matter." *Scientific American*, Vol. 288, No. 3, March 2003, pp. 50–59.

Cowen, Ron. "Dark Doings." *Science News*, Vol. 165, No. 21, May 22, 2004, pp. 330–332.

Cowen, Ron. "The Greatest Story Ever Told." *Science News*, Vol. 154, No. 25 and 26, December 19 & 26, 1998, p. 392.

Cowen, Ron. "Planet Signs? Sifting a Dusty Disk." *Science News,* Vol. 166, No. 15, October 9, 2004, p. 227.

Freedman, Wendy L. "On the Age of the Universe." *Daedalus*, Vol. 132, No. 1, Winter 2003, pp. 122–126.

Freeman, Ken C. "The Hunt for Dark Matter in Galaxies." *Science*, Vol. 302, No. 5652, December 12, 2003, pp. 1902–1903.

Hester, J. Jeff, et al. "The Cradle of the Solar System." *Science*, Vol. 304, May 21, 2004, pp. 1,116–1,117.

Hu, Wayne, and Martin White. "The Cosmic Symphony." *Scientific American*, Vol. 290, No. 2, February 2004, pp. 44–53.

Krieger, Kim. "New Measurement of Stellar Fusion Makes Old Stars Even Older." *Science*, Vol. 304, No. 5675, May 28, 2004, p. 1,226.

Perlmutter, Saul. "Supernovae, Dark Energy, and the Accelerating Universe." *Physics Today*, April 2003, pp. 53–60.

Pettini, Max. "Distant Elements of Surprise." *Physics World*, Vol. 16, No. 7, July 2003, p. 19.

Seife, Charles. "Illuminating the Dark Universe." *Science*, Vol. 302, No. 5653, December 19, 2003, pp. 2,038–2,039.

Simcoe, Robert A. "The Cosmic Web." *American Scientist*, Vol. 92, No. 1, January–February 2004, p. 30.

Veneziano, Gabriele. "The Myth of the Beginning of Time." *Scientific American*, Vol. 290, No. 5, May 2004, pp. 54–65.

Index

A

accelerators, 15–16, 18, 24,
48–50, 80
age of universe, 33, 48, 50, 54,
96–97, 102
ages of stars, 33, 48–51
antiproton, 130

B

baryon, 7–8, 109–111, 145, 149
Beta Pictoris, 182–185
big bang theory
ekpyrotic scenario, 52,
66–71
evidence for, 5–9, 27–28,
38–39, 102
pre–big bang scenario, 51,
64–66, 67–71
radiation and, 38–39
shortcomings of, 39–41
big bang nucleosynthesis
(BBN), 7–8
black holes, 15, 18, 23, 52, 54,
65–66, 135, 152–153
Branch, David, 77–78, 82

C

Caldwell, Robert, 121–122, 131
Carroll, Sean, 128–129
Cepheid, 6, 31–33
Chandra X-ray Observatory,
117–119
Cosmic Background Explorer
(COBE) satellite, 41, 102

cosmic microwave
background (CMB)
dark matter and, 148
discovery of, 8, 38–39, 93,
95, 102
inflation theory and, 21, 100,
105–107, 114–116, 118
measurement of, 73–74,
102–103
origin of, 5
sound and, 99–117
uniformity of, 8–9, 40–41,
101, 158
cosmological constant, 34, 46,
84, 85–87, 88, 90, 112, 124,
128–129

D

DAMA, 141, 143
dark energy
definition of, 14, 86
density of, 124, 129
and expansion, 87–89, 90, 113,
123–124
evidence of, 35, 73, 93–99,
111–112, 119, 125
weak lensing and, 125–126
dark matter
cold (CDM), 43, 46, 109–110,
111, 112, 136, 137, 145–146
composition of, 22, 132–144,
151, 159
dark halos and, 144–149, 164
detection of, 132, 138–149

Index

evidence of, 94, 97, 133, 135, 160–161
weak lensing and, 145, 148–149
deuterium, 8, 42, 170
Dirichlet membrane (D-brane), 63, 67
Doppler effect, 29, 153
Dvali, Gia, 129–130

E

Einstein, Albert, 13, 15, 16, 23, 62, 84, 94, 124, 128
dark energy and, 112, 115
"greatest blunder," 34–35
theory of gravity, 15, 16, 123, 131
theory of relativity, 5, 7, 27, 43, 44, 54, 57–58, 73, 110, 121, 129

F

fusion, 7–8
stellar, 49–51

G

graviton, 59, 130–131

H

Hubble, Edwin, 6, 9, 26, 27, 34, 73, 75, 84
Hubble's constant, 26, 28–29, 32–33, 35, 78
Hubble Space Telescope, 26, 28, 31, 32, 82, 90, 117, 178, 179

I

inflation theory, 14–15, 21, 40–41, 51, 56, 64, 70, 74, 83, 103, 115
evidence for, 100, 107, 114, 117–120
horizon problem and, 105
testing, 98, 116
inflaton, 56, 86, 112, 114–115
interdisciplinary approach, 11–12, 19–20, 24–25, 134

K

K-correction problem, 80, 81

L

Lykken, Joseph, 123, 131

N

neutralino, 136–143
neutrinos, 12, 15, 22, 23–24, 121, 127–128, 134, 135, 136, 138
1998 debate, 36–47
nucleosynthesis, 7, 169

O

Okamoto, Yoshiko K., 183–185
1a supernovas, 44, 72–91, 110, 117, 123–124

P

Peebles, Jim, 37–38, 45–46, 47, 103, 158
Penzias, Arno, 8, 28, 39, 102
Perlmutter, Saul, 9, 72–73
Pettini, Max, 166, 169–170
Phillips, Mark, 77, 82
power spectrum, 107, 108
Prochaska, Jason, 173, 174–175, 176
protons, stability of, 16, 22

Q

quarks, 11–13, 16, 58
quasars, 152–156, 163, 164, 169, 172–173

R

redshift, 6, 29, 62, 72, 74–75, 80, 81, 82–84, 87, 90, 91, 117, 119, 153–154, 172–173, 175–176

S

Sachs-Wolfe effect, 97, 111, 113–114

Sargent, Wallace, 154, 167
shape of universe, 41–43, 45, 52, 87, 97, 103, 108–109
Sloan Digital Sky Survey (SDSS), 97–99, 126, 162
solar system, origin of, 10, 176–181, 185
space-time, 5–6, 16–17, 21, 41, 54, 60, 63, 66, 98, 115, 126, 129–130
stars and origin of elements, 17, 169–176
string theory, 51–71, 144
Sunyaev-Zel'dovich effect, 115–116, 126
SuperNova/Acceleration Probe (SNAP), 91, 124–125, 127

T

T-duality, 61–63, 64
time reversal, 64

Turner, Michael S., 37–38, 42–43, 44, 45, 46–47, 118
2dF Galaxy Redshift Survey, 148, 162

W

weak lensing, 126–127, 145, 148–149
Weinberger, Alycia J., 184–185
Wilkinson Microwave Anisotropy Probe (WMAP), 50, 93–99, 102–103, 113, 116, 119, 148
Wilson, Robert, 8, 28, 39, 102

X

X-ray, 43, 117–119, 125–126

Z

Zel'dovich, Yakov B., 103, 158
Zwicky, Fritz, 75, 132, 159

About the Editor

Rick Adair has sailed research cruises in the Gulf of Alaska and the South Pacific Ocean, studied earthquakes, hydraulic fractures and geothermal wells, and helped collect data from the Mars Orbiter Camera, which still circles the Red Planet. He holds degrees in earth sciences from the University of California, Berkeley (BA) and from U.C. San Diego's Scripps Institution of Oceanography (Ph.D.).

Photo Credits

Front cover (top inset) © Royalty-Free/Corbis; (center left inset) © Digital Vision/Getty Images; (bottom right inset) R. Williams (STScI), the Hubble Deep Field Team and NASA; (bottom left inset) © Library of Congress Prints and Photographs Division; (background) Brand X Pictures/Getty Images. Back cover (top) © Photodisc Green/Getty Images; (bottom) © Digital Vision/Getty Images.

Designer: Geri Fletcher; Editor: Leigh Ann Cobb